CB Radio
Antennas

by David E. Hicks

HOWARD W. SAMS & CO., INC.
THE BOBBS-MERRILL CO., INC.
INDIANAPOLIS · KANSAS CITY · NEW YORK

THIRD EDITION

SECOND PRINTING—1975

International Standard Book Number: 0-672-21100-9
Library of Congress Catalog Card Number: 74-79355

Preface

The antenna, as simple as it may sometimes appear, is a vital component of any two-way radio system. The efficiency of the antenna determines, to a great extent, the distance over which radio signals can be transmitted and received.

At one time the antenna was considered relatively unimportant, and almost anything from a screen door to a random length of clothesline was used to serve the purpose. Because of the inefficiency of such an antenna, transmitter power had to be relatively high in order to achieve the desired communicating range. Design improvements over the years have resulted in antennas that provide communications over great distances while requiring only a fraction of the transmitter power needed previously. Because of the low rf power limitations imposed by the FCC on CB transmitters, the design and efficiency of the antennas used in this service are of particular importance.

This book was written to acquaint you with the importance of a good antenna system and to familiarize you with the various types of CB antennas and their theory of operation. You can learn what communicating range to expect, how radio waves react under various conditions, how to properly select and install base and mobile antennas, and how to improve the efficiency of existing antenna systems. You don't have to be an electronics engineer to understand this book. On the other hand, this book is not so basic that it will insult the intelligence of those with a more advanced knowledge of the subject. It's designed specifically to help those planning to set up a station and to aid those desiring to test, service, or improve their present system. If these goals are accomplished, this book has achieved its purpose.

DAVID E. HICKS

To Toni, Sharon, Brenda and Jack . . . my wife and children, who unselfishly sacrificed many days of companionship to help make this book possible.

Contents

CHAPTER 6

CHAPTER 7

CHAPTER 8

CHAPTER **1**

The Antenna System

Antennas are a vital part of radio. Some form of antenna, no matter how simple or complex, is required to effectively radiate and receive radio signals. Just how efficiently this is done, however, depends greatly on the performance of the antenna system.

IMPORTANCE OF A GOOD ANTENNA

In the early days of radio, antennas were generally crude and very inefficient. It was not uncommon to find a metal clothesline, fence, or even a screen door being used to transmit and receive radio signals. While such crude antennas did permit communication of sorts, they left much to be desired. Because they did not perform efficiently (make the most of the radio energy they were required to handle), the transmitter power had to be relatively high in order to achieve a satisfactory communicating range.

As radio developed over the years, it was found that through careful antenna design, the communicating range could be increased without increasing transmitter power. Design improvements resulted in even greater efficiency and led to the current models which permit radiocommunications over surprisingly long distances with only a fraction of the transmitter power formerly required.

With CB (Citizens-band) equipment, antenna performance is particularly important because of the low-power limitations imposed by the FCC (Federal Communications Commission). The maximum, legal power input permitted class-C* and -D CB stations is 5 watts. This is the input power to the final stage of the transmitter—not the output power. In actual practice, a CB transmitter with 5 watts of input will generally deliver no more than 3.5 watts of radio energy to feed the antenna. The output may even be less, depending on the efficiency of the transmitter, circuit adjustments, etc. Quite often the input is much less than 5 watts, in which case the transmitter output may be only a fraction of 1 watt. This, of course, means we will be dealing with relatively weak signals, compared with those originating from stations in other radio services. The Citizens Radio Service is designed to permit short-distance communications, and for this reason the transmitter power is limited.

However, by careful selection and installation of an antenna, it is possible to utilize to the fullest what little power is permitted and thereby obtain optimum performance from the radio equipment. In order to do this, the antenna must be capable of both radiating and receiving as much radio energy as possible, and this can be done only when minimum losses are incurred within the system. A two-way radio system is analogous to a chain, which is only as strong as its weakest link. Regardless of how powerful the transmitter or how sensitive the receiver, the system is only as good as its antenna.

With a good antenna it is possible to overcome, at least partially, some of the apparent shortcomings of CB radio. One of the biggest problems is limited ground-wave range, due in part to the low transmitter power (class-A stations excepted) used in CB operation. Assuming the transmitter is already operating at the maximum, legal power input and is properly adjusted, then only the antenna remains as a means of providing the additional radio energy needed to extend the range. With the transmitter power held constant, improving the efficiency of the antenna will increase the

---

* A maximum power input of 30 watts is permitted on one channel (27.255 MHz) in this class of service.

effective radiated power and thus extend the range of communications. This can be accomplished by reducing the amount of signal loss (increasing the amount of usable energy) in the existing antenna system or by installing a more efficient antenna.

Simple vertical antennas—those consisting strictly of a vertical radiating element—can utilize only that amount of energy being fed to them. Thus, if a transmitter having an output of 2.5 watts is connected to such an antenna, the most power that can be expected from it, even assuming no loss, is 2.5 watts.

A "gain" antenna, however, is capable of concentrating radio energy in such a way that it will appear much stronger than the signal producing it. By using gain antennas, it is possible to make 3 watts from the transmitter "look" like 30 watts when leaving the antenna. This means that the transmitter power will in effect be increased 10 times, making it equivalent in performance to a 30-watt transmitter. Thus, the extended range and improvement in overall performance that could be realized with a 30-watt transmitter can be achieved with only 5 watts by using a suitable antenna. In fact, even in those radio services where transmitter power can be increased, more "watts per dollar" (in effective radiated power) can often be obtained by improving the antenna system.

In considering the efficiency of an antenna, the entire antenna system must be taken into account. This includes not only the antenna itself but also the transmission line and any loading, matching, or coupling devices ahead of or following it. The inability of any one of these components to do its job efficiently can have adverse effects on the performance of the entire system.

It is not always possible to select the type of antenna to be used with CB equipment. Quite often the antenna is a physical part of the radio itself, as in the case of the many hand-held transceivers which employ built-in telescopic whip antennas similar to the one in Fig. 1-1. These transceivers are designed for close-range portable operation and are powered by self-contained batteries. Transmitter input power generally is less than 5 watts, and the adjustable antenna and its associated circuitry are designed to provide the best performance possible. The larger mobile and base-station transceivers that operate at or near the full 5-watt input are usually designed to use a remote (outside) antenna.

DESIGN CONSIDERATIONS

The antenna design determines its performance characteristics and, in turn, its usefulness. Furthermore, the physical as well as the electrical aspects of the antenna must be taken into consideration since both are interrelated.

Fig. 1-1. A class-D transceiver with a built-in telescopic antenna.

Obtaining the Desired Operating Characteristics

Some antennas are designed for general use, while others are made to exhibit certain characteristics that make them more suitable for a specific type of operation. Antennas can be made to radiate and receive radio signals equally well in all directions (omnidirectional) or to respond more in a given direction (unidirectional). Both types have advantages and disadvantages. Omnidirectional antennas are used most frequently in base and mobile installations where communications are desired in all directions at the same time. Unidirectional antennas, on the other hand, are better suited for point-to-point communications. They, too, can be used for communicating with stations in all directions; but, because of their radiating characteristics, they must be turned toward the desired station.

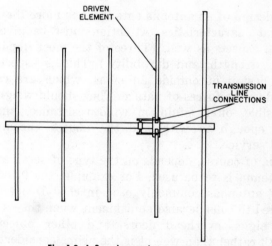

Fig. 1-2. A five-element beam antenna.

An antenna may consist of nothing more than a single element, or it may be more complex, having a number of elements. In either instance, only a single element (sometimes in two or more pieces) is used to radiate and receive radio signals. In the five-element antenna of Fig. 1-2, the driven element (the one to which the feedline connects) is the only one that transfers radio energy to and from the equipment. It is the number of elements and their length, spacing, and relationship to each other that determine the operating characteristics. Each element has a specific function and contributes in some way to the overall performance of the antenna. The driven element, for example, cannot be just any length; it must have a specific mathematical relationship to the wavelength of the radio signals it radiates or receives. If its length is incorrect, a signal loss will occur. The rule is that the higher the frequency, the shorter the elements.

Through proper design, an antenna can be made to provide signal gain, send radio energy in all directions or concentrate it only in one, and direct this energy along the surface of the earth, toward the sky, or both. These are but a few of the effects that can be achieved; they are the most important, however.

Construction

The design of an antenna entails much more than just the electrical characteristics. Attention must be given to the physical aspects as well. (Three of the most important are weight, strength, and durability.) This is especially true in the design of outside antennas, which are constantly exposed to the forces of nature. They should weigh as little as possible, offer minimum wind resistance, and yet be sturdy enough to withstand the abuses encountered in normal service.

Much, of course, depends on the type of service in which the antenna is to be used. For example, the 9-foot mobile "whip" antennas commonly used in class-D operation are subjected to considerable punishment when they strike tree limbs, signs, overhead doors, and other objects. Here, strength rather than weight is the prime consideration; for this reason, such antennas are generally made of hardened steel or fiber glass. Base-station antennas, on the other hand, usually consist of several elements and are mounted at greater heights. Here, weight must be kept as low as possible but not at the expense of strength. The elements are usually made of hollow aluminum tubing—which is light in weight, yet sturdy enough to withstand high winds. Obviously, the lighter the antenna, the easier it will be to install—or to take down should servicing be required. At the same time, when you consider the length of the elements on class-D antennas and the additional strain imposed by wind, ice accumulation, etc., it is apparent that strength is also a necessity. Durability, of course, is important with all types of antennas.

Antenna Switching

An antenna that is suitable for radiating rf energy is equally capable of receiving radio signals. In fact, the transmitting and receiving characteristics of a given antenna are essentially the same. For this reason, practically all two-way radio equipment, including CB radio, uses the same antenna for both transmitting and receiving.

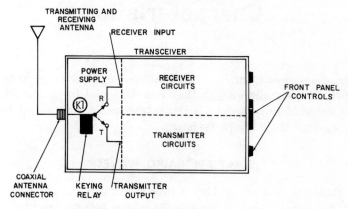

Fig. 1-3. Basic antenna-switching arrangement.

A switching arrangement, such as that shown in Fig. 1-3, enables the antenna to be connected to either the receiver or the transmitter. For the purpose of illustration, the transceiver is shown divided into sections. The transmitter output and receiver input circuits are terminated at keying relay K1. When this relay is in a resting, or de-energized condition, the antenna is connected (through its contacts) to the receiver input circuit as shown. When the microphone is keyed to transmit, however, relay K1 becomes energized and switches the contacts from the receive (R) position to the transmit (T) position. In this position, the antenna is connected to the transmitter output circuit.

CHAPTER 2

Radio Wave Characteristics

A clear understanding of antennas requires some knowledge of the radio energy they are designed to handle. The behavior of radio waves is important and a number of factors affect their characteristics.

WHAT ARE RADIO WAVES?

Radio waves are not of a physical nature; that is, they cannot be seen or felt but instead are recognized only by their effects. Consider the magnetic field that surrounds a permanent magnet (Fig. 2-1). This energy, too, cannot be seen or felt, yet its presence is apparent from the attracting force it exerts on certain kinds of metal. Nobody knows exactly what a magnetic field consists of. Theoretically it is considered to be made up of "lines of force." These lines form a field pattern around a magnet and by the direction in which they lie, indicate which way the force they exert will act on an object. The strength of a magnetic field is dependent on the density of the lines of force—the more lines there are, the stronger the field that will be produced.

Although magnetic energy itself cannot be seen, it is possible to obtain a "picture" of the field pattern by means of iron filings. When sprinkled over the surface of a sheet of

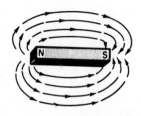

Fig. 2-1. The magnetic field produced by
a permanent magnet.

paper placed in contact with a magnet, the iron particles will assume positions in accordance with the lines of force in the magnetic field. Fig. 2-2 shows the pattern that was obtained when two bar magnets were positioned to attract each other.

A permanent magnet is not the only source of magnetic energy. From Fig. 2-3 it can be seen that electric current flowing through a conductor also produces magnetic lines of force. These lines are always at right angles to the conductor and their direction (indicated by the arrows) is determined by which way the current is flowing through the conductor. Furthermore, the number of magnetic lines and the resultant energy around the conductor is determined by how much current is flowing through it. From this we see that a direct relationship exists between electric and magnetic energy.

In radio we are concerned with three kinds of fields—electric, magnetic, and a combination of the two known as electromagnetic. A static magnetic field is one that is stationary and has constant intensity, such as that produced by a permanent magnet. Likewise, a static electric field is produced by a stationary electric charge of constant intensity. Now, should either an electric or a magnetic field change in intensity or move through space, such change or movement will produce the opposite field at right angles.

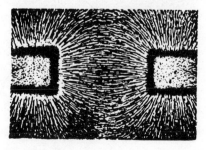

Fig. 2-2. A magnetic field pattern obtained with iron filings—clearly indicating the lines of force.

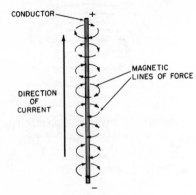

Fig. 2-3. Magnetic lines of force produced by current flowing through a semiconductor.

In other words, a moving electric field or one that is varying in intensity will produce a magnetic field. It is this principle that makes radiocommunications possible.

Radio waves are produced by an ac (alternating current) which changes in intensity at a very rapid rate. Alternating current, just as its name implies, flows first in one direction and then in the other. This change occurs at a predetermined rate, or frequency. Alternating current can be visualized as a row of marbles within a hollow tube representing a conductor (Fig. 2-4). The marbles, like alternating current,

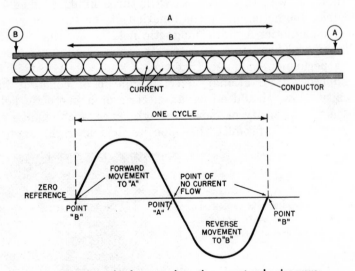

Fig. 2-4. Relationship between alternating current and a sine wave.

move back and forth, but no marble actually travels from one end of the tube to the other. That is, while their movement is transferred throughout the length of the tube, the actual distance traveled by any marble in either direction is relatively short. In this illustration, the marbles will first move to point *A* and stop; then reverse direction and travel to point *B*. This action can be plotted as a sine wave, as shown in Fig. 2-4. Here, the forward movement to *A* is represented by that portion of the wave above the reference line, and the reverse movement to *B* by the other half appearing below the reference line. The initial forward movement is indicated by the rise of the curve above the reference line. The drop off beyond the peak results from the slowdown and eventual stoppage at point *A*. Movement then begins in the reverse direction (indicated by that portion of the curve below the reference line), reaches maximum, and once again slows to an eventual stop—only now at

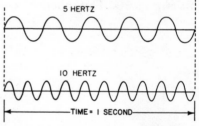

Fig. 2-5. Relationship between alternating current at 5 and 10 hertz.

point *B*. This sequence is termed a "cycle" and is expressed in hertz. Thus, the frequency of alternating current is determined by how many cycles occur each second. Fig. 2-5 shows a comparison between alternating currents visualized as sine waves at frequencies of 5 and 10 hertz.

A conductor carrying the standard 60-hertz household current will have a small field of energy surrounding it (Fig. 2-3) but will hardly provide the type of radiation needed for radiocommunications. Even if the power of 60-hertz ac were increased considerably, it would radiate only a few hundred feet. Perhaps you have heard the buzzing sound from the speaker of a car radio while driving under high-voltage power lines suspended across a highway. Radiation from these lines affects normal broadcast reception for only a few hundred feet and is most noticeable when the radio is tuned to a weak station. Obviously, then, even the radiation from high-powered 60-hertz ac is not suitable for radiocommunications.

The frequency of the current needed to produce radio waves is in the thousands and even millions of hertz. For convenience, the prefixes kilo (meaning thousand) and mega (meaning million) are used in radio. A frequency of 1 kHz (kilohertz) equals 1000 hertz, and 1 MHz (megahertz) equals 1000 kilohertz, or 1,000,000 hertz.

Skin Effect

Electric current at these high frequencies exhibits some unusual characteristics. For example, alternating current at the standard 60-hertz power-line frequency flows in equal amounts throughout the entire cross-sectional area of a wire, as shown in Fig. 2-6A. At radio frequencies, however, the current tends to flow near the outer surface, as shown in Fig. 2-6B. This phenomenon is known as "skin effect."

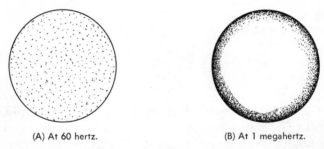

(A) At 60 hertz.   (B) At 1 megahertz.

Fig. 2-6. Cross-section of a conductor showing the effect of frequency on current flow.

Consequently, the center portion of a conductor carrying such high-frequency current contributes little toward its transfer. For this reason, hollow conductive tubing is often used in the output circuitry of radio transmitters.

Not only does this high-frequency current flow along the outer surface of a conductor, it also causes a form of energy to be radiated from it which, under the proper conditions, is capable of traveling thousands and even millions of miles in free space. This energy resists any efforts to confine it within a conductor; thus, by encouraging such radiation from a properly designed antenna, it can be utilized as a means of communications. The frequencies at which this radiated energy becomes useful for communications is known as radio frequencies, or simply rf. (One of the major factors that determines the behavior of radio waves is their frequency.) Fig. 2-7 shows the rf spectrum and the various bands into which it has been divided. All CB operation is within the hf (high frequency), vhf (very-high frequency) and uhf (ultrahigh frequency) bands. Specifically, the class-A CB channels are allocated in the uhf band, the class-C channels are located in the hf band along with the class-D channels and also in the vhf band. Radio waves behave quite differently at hf, vhf, and uhf frequencies, as you will see later in this chapter.

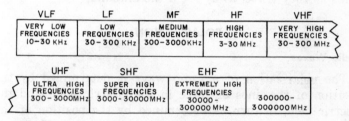

Fig. 2-7. The radio-frequency spectrum.

The rf current which produces radio waves is generated by the transmitter and fed to the antenna through a suitable coaxial cable. This cable has an outer shielding braid which, among other things, tends to confine the rf energy within the cable, thereby discouraging radiation from points other than the antenna itself.

Composition of Radio Waves

Radio waves, like light waves, are a form of electromagnetic energy composed of moving fields of electric and magnetic forces. The lines of force in the two fields are always at right angles to each other. That is, when the lines of force in the electric field are horizontal, those in the magnetic field will be vertical and vice versa. Fig. 2-8 shows the relationship between the lines of force within these two fields. The arrows on the lines indicate their direction (to be discussed shortly) at a given instant. The energy of radio

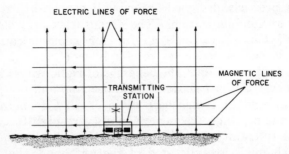

Fig. 2-8. Relationship between the lines of force in the electric and magnetic fields of a radio wave.

waves is contained within the electric and magnetic fields and is alternately passed back and forth between the two fields as they travel away from the antenna.

POLARIZATION

Radio Waves

The polarization of a radio wave is determined by the position of the lines of force in the electric field, and the electric field, in turn, is determined by the position of the radiating element of the transmitting antenna. Hence, if the electric lines of force run parallel to the ground, the radio wave is horizontally polarized; and if the lines of force run perpendicular to the ground, the wave is vertically polarized. Of course, in either instance the lines of force in the magnetic field will be at right angles to the electric lines of force. It follows, then, that a transmitting antenna with a horizontal radiating element will emit horizontally polarized waves, and that one with a vertical element will emit vertically polarized waves. Just remember it is the position of the radiating element (with respect to ground) that determines the polarization of the transmitted radio wave. The wavefront shown in Fig. 2-8 is vertically polarized, and the electric lines of force are exactly at right angles to the ground. They could, however, be at any angle with respect to ground and still maintain the same relationship to the lines of force in the magnetic field. Fig. 2-9 illustrates theoretically how a wavefront would appear when radiated from an antenna bent to the angle shown.

Polarization is important in two-way radio because it affects both the transmission and the reception of radio signals. For example, the polarization of radio waves has a considerable influence on their behavior as they travel away from the transmitting antenna, and it also dictates the type of receiving antenna needed for maximum signal pickup. For simplicity, consider only the electric lines of force, in connection with polarization. Also, it is well to remember that these lines are always parallel to the radiating ("driven") element of the antenna.

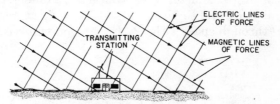

Fig. 2-9. Theoretical effect on the wavefront when the transmitting antenna is at an angle.

Antenna Polarization

When the driven element is positioned perpendicularly to the ground, it will radiate vertically polarized radio waves, and the antenna is likewise termed vertically polarized. By the same token, a horizontally polarized antenna has the radiating element positioned parallel to the ground and radiates horizontally polarized waves. Although this is the accepted terminology, it is often confusing to those unfamiliar with antennas. The fact that an antenna is vertically or horizontally polarized does not necessarily mean it has poles like those of a battery or a permanent magnet; it merely indicates the polarization of the radio waves that the antenna is intended to radiate or receive. The polarization of an antenna is determined by its position with respect to the ground. Thus, an antenna that is considered to be vertically polarized by virtue of its normal position can be made to act as a horizontally polarized antenna by merely reversing its position.

From this we see that polarization is not an inherent property of the antenna itself, but rather a matter of the position of the driven element with respect to ground. Some commercial antennas are designed for only vertical or horizontal polarization, whereas other antennas may be positioned to provide either vertical or horizontal polarization.

Usually a simple vertical or horizontal antenna consists of just a driven element, and its polarization is therefore obvious. With some of the more complex antennas, however, the polarization is not always readily apparent because of the increased number of elements and their relationship to each other.

The nature of radio waves is such that current will be induced in any conductor they contact. To understand this action, it is helpful to visualize a radio wave as a sine wave moving through space (Fig. 2-10). As you will recall, a sine

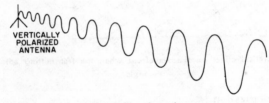

(A) Vertically polarized waves.

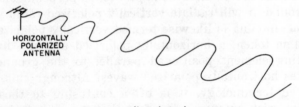

(B) Horizontally polarized waves.

Fig. 2-10. Radio energy pictured as sine waves traveling in space.

wave was used previously in Fig. 2-4 to indicate the movement of alternating current in a conductor. This type of wave can likewise represent changes in a radio wave, since these changes are actually caused by and in accordance with the rf current flowing in the transmitting antenna. Each half cycle of this sine wave represents a reversal in the direction of the lines of force in the electric and magnetic fields.

Every line of force, whether electric or magnetic, has a property known as "direction." This is the direction in which the force exerted by the line will act on an object. Referring to Fig. 2-8, you will notice that arrows are shown on the lines of force in this wavefront to indicate their direction at a particular instant. Actually the direction of these lines is continually reversing in step with the alternating current which produces them. In other words, by comparing the lines of force shown in Fig. 2-8 with one cycle of a sine wave shown previously in Fig. 2-4, the relationship between the two will become apparent. The direction (indicated by the arrows) of the lines of force in Fig. 2-8 at that instant is comparable to the forward movement to A of alternating current represented by the upper portion of the sine wave in Fig. 2-4. On the reverse half of this cycle (lower half of sine wave), the direction of the lines of force in both fields of the radio wave in Fig. 2-8 will be the opposite of that shown. When a conductor—a receiving antenna, for example—is exposed to a radio wave, a current is induced. The radio waves represented in Fig. 2-10 indicate the changing intensity of the electric field since the lines of force of this field determine the polarization; however, in the receiving antenna, current is induced only by the lines of force in the magnetic field. Furthermore, the induced current will be maximum when these lines are at right angles to the receiving element, and it will become less as this angle diminishes. The magnetic field, although not shown, changes in the same manner as the electric field but moves at right angles to it. The amount of current induced will depend on the field strength of the arriving wave, and the direction of the current in the antenna is dependent on the direction of the magnetic lines of force at that particular instant. Hence, the current induced in the receiving antenna is an alternating current and is nothing less than a "carbon copy" of that which originally produced the radio waves at the transmitting antenna. The amount of current flowing in the receiving antenna will naturally be less because of the energy dissipated as the radio waves move through space.

When we speak of the transmitting and receiving antennas of a station, this does not imply that separate antennas are employed. Rather, the reference is made only to the function of a single antenna during one or the other operating modes. Practically all commercial and CB two-way radio equipment is designed to use the same antenna for both transmitting and receiving, as was explained previously in Chapter 1.

The polarization of a receiving antenna has a considerable effect on the amount of current that radio waves will induce. Obviously, it is desirable that maximum energy be fed to the receiver circuits. Since the electric lines of force determine the polarization of radio waves, they also dictate the polarization that the receiving antenna must have to provide satisfactory reception. Theoretically, then, this would indicate that vertically polarized radio waves could be received only with a vertically polarized antenna, and horizontally polarized waves could be received only with a horizontally polarized antenna. In reality, this does not hold true because a wavefront tends to be broken up into components of varying polarization while en route to the receiving antenna. It is these variations in polarization that make it possible for current to be induced in a receiving antenna which is oppositely polarized from the transmitting antenna. Radio waves tend to maintain their polarization better at some frequencies than at others. Despite the variations of polarization in individual lines of force, the overall or predominating polarization (sun polarization) of the received wave can be considered the same as that of the transmitting antenna. In other words, for all practical purposes, maximum current will be induced in a receiving antenna when it has the same polarization as the antenna that is radiating the signals.

PROPAGATION

Until now we have been concerned primarily with the composition of radio waves and the manner in which they induce energy in the receiving antenna. Also of importance is the propagation of radio waves, for it has a considerable influence on the strength, reliability, and range of radio signals.

The first part of this discussion will be devoted to the propagation characteristics of radio waves in general. Later in this chapter we will more specifically analyze how these waves react at CB frequencies. In this way the problems associated with propagation at CB frequencies will be easier to understand.

After leaving the transmitting antenna, radio waves may be refracted (bent), reflected, broken up, or influenced in some other way as they travel through space. Frequency, type of antenna, earth, and atmosphere are just a few of the factors that affect propagation. In free space, radio waves travel at the speed of light (300,000,000 meters or approximately 186,000 miles per second). In other mediums, however, they travel somewhat slower. In fact, the velocity of radio waves in dielectrics is inversely proportional to the dielectric constant. When radio waves pass from one medium into another of a different dielectric constant, their velocity changes abruptly and the waves are refracted from their original course. (More about this later.)

Spreading

In radiating from the antenna, radio waves spread outward, occupying an ever-increasing area as shown in Fig. 2-11. The farther they travel, the less their field strength (energy) becomes. Under purely free-space conditions, the field intensity of radio waves would be inversely proportional to the distance from the transmitting antenna. In other words, the signals received at station B in Fig. 2-11 would be exactly half as strong as those at station A, since station B is twice as far from the transmitting station. In reality, the inverse-distance rule does not hold true because energy is absorbed from the waves by the earth and atmosphere.

The spreading of radio waves is comparable to the shock waves, or ripples, produced when a rock is dropped into the water. The ripples are quite large near the point where the rock hits the water, but become smaller and smaller as they move outward. Finally, the ripple becomes so weak it is no longer visible. The larger the rock, the farther the ripple is capable of traveling before being dissipated. Essentially the same is true of radio waves.

The radio energy leaving the transmitting antenna consists of ground waves and sky waves. Ground waves travel along the earth's surface and even follow its curvature to some degree. Sky waves, on the other hand, travel upward

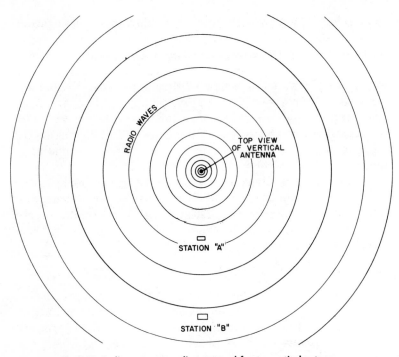

Fig. 2-11. Radio waves spreading outward from a vertical antenna.

through space. Both types can be further classified according to the manner in which they move along their respective paths, and their behavior is determined to a great extent by their frequency. For example, at lower radio frequencies, both ground and sky waves—although they behave quite differently—are generally capable of providing long-distance radiocommunications. At the higher frequencies, however, both ground- and sky-wave propagation change considerably and so does the communicating range.

Ground Waves

The earth and its atmosphere do not appear the same to all radio waves at all frequencies. Below approximately 5 MHz, the ground acts like a conductor to radio waves, but at higher frequencies, the ground assumes the characteristics of an insulator. Below 5 MHz, energy is induced in the ground by the radio waves as they move across its surface, and the entire wavefront tends to follow the curvature of the earth as seen in Fig. 2-12A. Ground-wave propagation at these frequencies will generally permit communications with stations located some distance beyond the horizon. Just how far these ground waves will travel depends on how fast their energy is dissipated. The amount of ground-wave attenuation per given distance is determined by such things as the frequency of the radio signals, resistivity of the ground, and terrain features over the signal path, to name just a few.

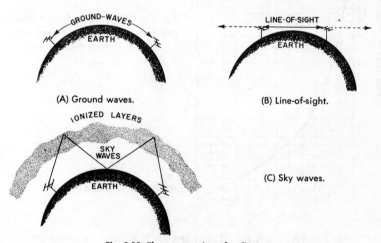

(A) Ground waves.

(B) Line-of-sight.

(C) Sky waves.

Fig. 2-12. The propagation of radio waves.

As their frequency is increased, radio waves follow the earth's curvature less and less. Eventually they travel straight outward from the horizon and into space, as shown in Fig. 2-12B. Since less surface area is covered, the ground-wave communicating range is much less here than in the previous example.

Frequency has a number of other effects on the propagation of radio waves. For example, ground waves at the lower frequencies maintain their polarization fairly well as they move across the earth's surface; whereas at higher frequencies, the wavefront tends to be broken up more readily into components of varying polarization. Furthermore, these waves do not bend and reflect as they do at lower frequencies, and as the frequency is increased, their energy is dissipated much faster. Fig. 2-13 will give you some idea

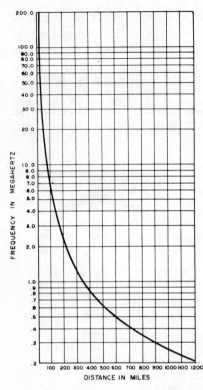

Fig,. 2-13. The relative effect of signal frequency on the ground-wave communicating range.

of the effect which frequency has on ground-wave propagation.

Sky Waves

As mentioned previously, that portion of the radiation which travels upward from the antenna into space is known as a sky wave. At high frequencies, the waves generally continue on through space until they are dissipated. However, at the lower frequencies (7 MHz, for example), sky waves may be reflected to earth a considerable distance away from their origin. This action is known as a "skip." It may occur only once (single skip) or several times (multiple skip), as shown in Fig. 2-12C. This is the type of propagation that makes worldwide radiocommunications possible.

Sky waves are refracted and reflected by ionized layers which surround the earth. For the purpose of illustration, only one layer was shown in Fig. 2-12C. In reality, there are several ionized layers ranging from 50 to 400 miles above the earth. These layers are designated as D, E, F_1, and F_2. The last two are actually one layer (F) which divides periodically and hence is subdivided into 1 and 2. Constant changes (some predictable, others not) in these ionized layers vary the propagation characteristics of sky waves considerably. Although these variations usually follow certain established patterns, on occasion they present some rather unusual effects. Fig. 2-14 shows the altitudes of the various ionic layers as well as their relationship to one another. You can see that they change not only from season to season but also between night and day.

As mentioned previously, when radio waves pass from one medium to another having a different dielectric constant, their velocity changes abruptly and they are reflected (bent) from their original course. As a result, sky waves passing through the ionized layers will be bent but to just what extent will depend on their frequency. They may merely be diverted—or, through a series of refractions, they may be reflected back to earth.

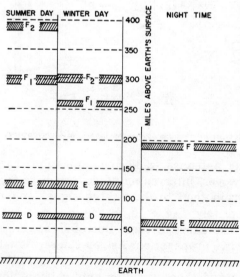

Fig. 2-14. Relationship between the ionic layers during winter and summer and night and day.

Propagation at CB Frequencies

Now let us consider ground- and sky-wave propagation at CB frequencies, and compare it with what has been discussed so far.

All class-A CB operation is within the uhf band of the spectrum. Here, radio waves act somewhat like a beam of light. That is, they follow a more or less straight path from the antenna and are attenuated rapidly by intervening objects along the ground-wave paths. Theoretically, uhf radio waves afford line-of-sight communications only (see Fig. 2-12B). With no obstructions between the transmitting

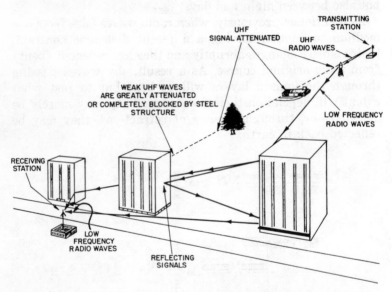

Fig. 2-15. The effect of intervening objects on high- and low-frequency radio waves.

and receiving antennas, radiocommunications can be very reliable. However, intervening objects such as trees, buildings, or terrain tend to diffuse the signal and may even block it completely. Fig. 2-15, although somewhat exaggerated, shows the effect of these line-of-sight characteristics on ground-wave propagation. As you can see, low-frequency radio signals are reflected from buildings and other objects quite readily and thereby suffer only a slight loss by the time they reach the receiving antenna. The uhf signal, however, follows a more or less straight (line-of-sight) path and is greatly attenuated by intervening objects; the amount of attenuation depends on the number of objects and their composition. Terrain and large steel buildings account for most of the signal attenuation; terrain alone may block the signals completely. Radio energy is dissipated quite rapidly at uhf frequencies; hence, the communicating range is rather limited.

Ground-wave propagation is much better at the class-D CB frequencies (26.965 to 27.255 MHz). Here, the much lower frequencies overcome to some degree the line-of-sight

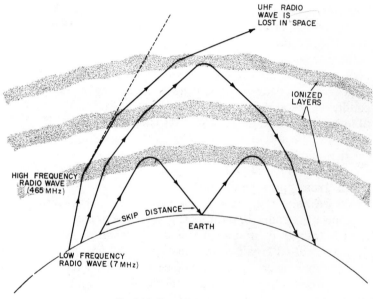

Fig. 2-16. Sky-wave propagation.

characteristics of uhf signals. Still, a direct path between antennas produces the best results; but because of the relatively lower frequencies, the ground waves are not attenuated nearly as much and therefore provide more-reliable communications over a somewhat larger area.

Frequency also has a great effect on sky-wave propagation, as seen in Fig. 2-16. Radio waves traveling upward through one medium are refracted as they pass through the ionized layers where the propagation velocity is less.

Depending on how much these waves are refracted—and this is determined by their frequency—they may or may not return to earth. The higher the frequency, the less the refraction of the radio waves. This can be seen by comparing the paths taken by the waves in Fig. 2-16. The lower-frequency wave is progressively refracted by the ionized layers and finally reaches an angle where it is reflected back to earth. However, the higher-frequency wave tends to travel straight through the layers, or at best is refracted only slightly. Obviously, that portion of the energy radiated as sky waves is of little use at class-A CB frequencies. The same is true at class-D frequencies under normal conditions. For this reason, most CB antennas are designed to concentrate, along ground-wave paths, the energy that would normally be lost as sky waves.

Sky waves at the class-D frequencies are not always lost, however. Occasionally, masses of relatively dense ionization appear at or near the same altitude as the E layer. This condition, referred to as sporadic E ionization, can cause abnormal refraction of sky waves at frequencies up to 30 MHz—and even at vhf frequencies, depending on the intensity of the ionization. When this condition exists, radio signals at class-D CB frequencies may be reflected back to earth from these ionized masses in the same manner as radio waves at the lower frequencies. When this skip condition occurs, radiocommunications between stations separated by hundreds and even thousands of miles are possible. These so-called freak conditions generally occur during the spring and early summer. On occasion, extended ground-wave propagation will also increase the communicating range beyond the normal expectations.

Although CB communications are limited entirely to ground waves, the changes in ionized layers that affect sky-wave propagation do not always follow their established patterns. Therefore, while a comprehensive discussion of sky-wave propagation may seem somewhat unnecessary in connection with CB radio, it can be quite helpful in recognizing and understanding some of these unusual events when they occur.

CHAPTER **3**

Antenna Characteristics

Since practically all CB two-way radio equipment uses the same antenna for both transmitting and receiving, there are more factors to be considered than if the antenna were used only for receiving. With two-way radio antennas we will be concerned with such things as gain, front-to-back ratio, impedance, standing waves, and line losses, to name just a few. All or only part of these may apply in any particular instance. Much depends, of course, on the type of antenna. Front-to-back ratio, for example, pertains only to directional antennas, while impedance and line losses are important with any type. In the pages that follow, we will go into more detail on the physical and electrical characteristics of antennas and discuss the various terms associated with them.

ANTENNA LENGTH

A direct relationship exists between the physical and electrical characteristics of an antenna. For example, an antenna chosen with no particular length in mind will not exhibit the same characteristics at all frequencies. Instead, it will respond efficiently only when its length has a specific mathematical relationship to the wavelength of the radio signals applied to it.

To understand what is meant by wavelength, refer to Fig. 3-1. By dividing the velocity of a radio wave by its frequency (number of cycles per second), we are able to determine its wavelength, or the distance it will travel during the time required for one cycle to occur. A radio wave, then, with a frequency of 27 MHz will travel a distance of 11 meters during one cycle. A meter is a unit of measurement equal to 39.37 inches. Thus, by multiplying 39.37 by 11 you will find that the length of a radio wave at 27 MHz is 433.07 inches, or over 36 feet long. The length of any radio wave can be found by the formula:

$$\lambda = \frac{300,000}{f}$$

where,
  λ is the wavelength in meters,
  f is the frequency in kilohertz.

This is the same as:

$$\lambda = \frac{300}{f}$$

where,
  λ is the wavelength in meters,
  f is the frequency in megahertz.

The term "meter" is used quite often in two-way radio. It is quite common to refer to the wavelength of radio waves rather than to their frequency. Amateur (ham) radio

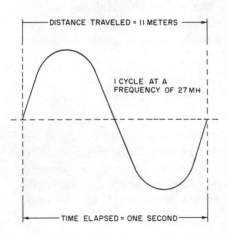

Fig. 3-1. A sine wave representing one wavelength at 27 MHz.

is a good example of this usage. A ham would be more likely to say that he operates on 40 meters rather than on 7 MHz, although they are one and the same. The class-C and -D CB frequencies occupy what was formerly the 11-meter amateur band, and this band is still commonly referred to as 11 meters. As the frequency of a radio wave is increased, its wavelength is decreased. In other words, if the frequency of the radio wave shown in Fig. 3-1 is doubled (2×27 MHz = 54 MHz), two cycles rather than one will occur over a distance of 11 meters. This means the wavelength at 54 MHz will be exactly half that at 27 MHz. Therefore, as the order of meters decreases, the frequency increases. At 10 meters, for example, the frequency is higher (wavelength shorter) than at 11 meters; hence the common expression "short wave," referring to high-frequency radio signals.

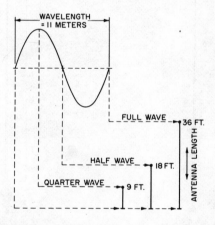

Fig. 3-2. Relationship between wavelength and antenna length.

An antenna will not operate efficiently at any given frequency unless its physical or electrical length is the same as, or a multiple of, the wavelength at that particular frequency. An antenna may be a wavelength long, or it may be a multiple such as a half or a quarter wavelength, as shown in Fig. 3-2. As you can see, a wavelength at 27 MHz (11 meters) is approximately 36 feet long. In reality it is somewhat longer than this; however, for simplicity we will stick to round figures. Although a full-wave antenna is very efficient, its length obviously makes it impractical in most cases for CB operation. On the other hand, an antenna cut to a half wavelength is essentially as efficient as the full wave, yet is only half as long, or approximately 18 feet. While a great many base stations employ half-wave antennas, their length would still make them rather precarious for mobile work and very impractical for portable operation. A quarter-wave antenna, however, is only 9 feet (108 inches) long and is therefore much better suited for mobile operation. This length of 108 inches is the overall length, as shown in Fig. 3-3A. When a spring mount is used, it, too, is considered part of the radiating element; so the rod itself should be only 102 inches long (Fig. 3-3B).

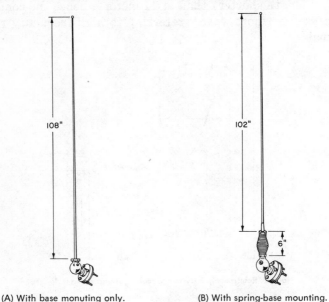

(A) With base monuting only.   (B) With spring-base mounting.

Fig. 3-3. The length of a quarter-wave whip at 27 MHz.

It should be pointed out here that there is a difference between the physical and electrical length of an antenna. For example, the actual length of a half-wave antenna will not be exactly equal to a half wave in space but will be somewhat less. This difference is due to several factors, one being the thickness of the antenna element(s) in relation to the frequency of the radio waves. For all practical purposes, the thinner the element, the less the difference between the electrical and the physical length. This means, then, that an antenna with a small cross-sectional area will be longer electrically than one of the same length having a larger area.

Loading Coils

Quite often the length of an antenna presents problems, especially in portable and mobile operation. This is seldom true with class-A equipment because of the higher operating frequencies (hence, shorter antennas). However, it is sometimes a problem in class-D operation because even a quarter-wave antenna will be 9 feet long. Although quite efficient, this antenna is sometimes considered to be too long and

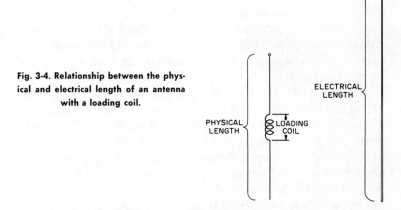

Fig. 3-4. Relationship between the physical and electrical length of an antenna with a loading coil.

PHYSICAL LENGTH

LOADING COIL

ELECTRICAL LENGTH

is cut to less than a quarter wavelength so that it will be more manageable. When the antenna is shortened, its efficiency is reduced. By means of a loading coil, however, an antenna can be made to "appear" longer electrically than it actually is physically (Fig. 3-4). In this way, a shorter antenna can be used without reducing its efficiency. To be more specific, an ideal antenna is one that is purely resistive in nature. This characteristic is not always exhibited, however, because a certain amount of capacity often exists. A loading coil is an inductance especially designed to cancel out, or at least minimize, the capacitive effect of the antenna system. When an antenna is shorter than an exact quarter wavelength, it will present a capacitive reactance to the source of the signal. The more the antenna is shortened, the higher the capacitive reactance becomes; as a result, its radiation resistance is lowered. Maximum rf energy will be transferred from the transmitter to the antenna when the latter presents the proper load impedance to the transmitter. Hence, a reduction in the radiation resistance of the antenna will cause a similar reduction in output. By using a suitable loading coil to balance out the capacitive reactance of a shortened antenna, it is possible to present a much better load to the transmitter, thereby increasing antenna efficiency and hence signal output. Even with a loading coil, a shortened antenna may not be as efficient as one that is a full quarter wavelength. The longer antenna is usually more efficient because it has a higher radiation resistance and effective elevation; consequently, it can provide a somewhat greater communicating range. However, in certain cases the "shortened" antenna may be physically the most practical. For example, using a 9-foot antenna for portable operation, either indoors or outdoors, would be rather awkward to say the least.

Loading coils are used with a great many class-D (27 MHz) antennas designed for mobile and portable operation. To perform its intended function, the loading coil must be connected in series with the antenna. However, there seems to be a considerable difference of opinion as to where it should be placed for most efficient operation. As a result, we find that the loading coil is connected at the base of some antennas, while on others it may be located in the center, at the top, or even wound in a spiral along its entire length. Such antennas are generally referred to as *base-loaded* or *center-loaded* antennas, depending on the location of the coil. Loading coils themselves appear in a number of different forms and may or may not be a physical part of the antenna. Their design depends a great deal on their electrical characteristics, the type of antenna with which they are used, and the type of operation (indoors, outdoors, portable, mobile, etc.) intended.

Many antennas used with class-D portable equipment employ loading coils; however, their construction often differs considerably from the coils used with mobile antennas. One such example is the base-loaded whip in Fig. 3-5. This antenna is only 40 inches long. Its loading coil, which is covered with a vinyl jacket, is a physical part of the antenna itself, and unlike most loading coils, it is of an open construction. In comparison, Fig. 3-6 shows a base-loaded antenna, approximately 43 inches long, designed primarily for mobile roof-top installation. The loading coil is 4 inches long and 1 inch in diameter; the whip itself is 39 inches long. An example of a center-loaded antenna is shown in Fig. 3-7. This antenna is adjustable to a minimum length of 43 inches and an extended length of 60 inches. A top-loaded antenna is shown in Fig. 3-8. This mobile whip has an overall length of 50 inches. The polyethylene-enclosed coil capsule, which is permanently fused to the top of the stainless-steel whip, is completely impervious to climatic conditions. The entire loading coil is 10 inches long and approximately ¼ inch in diameter.

The current distribution in an antenna is affected to some degree by the insertion of a loading coil. That is, the current along the length of the antenna is not the same at all points. Fig. 3-9A, B, and C show the effect of base, center, and top loading, respectively, on the current distribution. There are many pros and cons with respect to the position of the loading coil in an antenna. Therefore, it will suffice to say that a loading coil may have advantages as well as disadvantages at any of these positions.

Fig. 3-5. The loading coil of this 27-MHz portable whip is physically a part of the antenna itself.

An antenna using a fourth method of loading, and one that could basically be considered a combination of the three methods of loading discussed so far, is shown in Fig. 3-10. This method, known as continuous loading, involves winding the coil spirally along the entire length of the antenna from bottom to top. The mobile antenna shown here is constructed of fiber glass and is only 3 feet long. Many CB antennas are constructed of this material, including some of the long (full quarter-wavelength) antennas

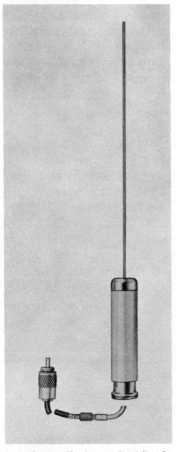

Courtesy The Antenna Specialists Co.

Fig. 3-6. This mobile whip (Model M-67) employs a base-loading coil.

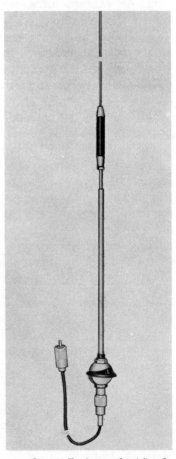

Courtesy The Antenna Specialists Co.

Fig. 3-7. The Model M-49 is an example of a 27-MHz mobile whip using center loading.

Fig. 3-8. A top-loaded mobile whip (Model TWL-M) designed for class-D operation.

Courtesy Hy-Gain Electronics Corp.

and particularly those antennas designed for marine applications.

ANTENNA GAIN

Although we speak of an antenna as having gain, this should not be misinterpreted as meaning that it can actually amplify radio signals like an electronic circuit. No antenna can provide gain in that sense. However, through proper design, an antenna can be made to concentrate its radiated energy in such a way that it will appear to have been produced by a much stronger source. For example, an antenna that concentrates most of the radio energy in one direction will provide an increase in coverage in that direction over another antenna which is fed the same amount of power but radiates energy equally well in all directions.

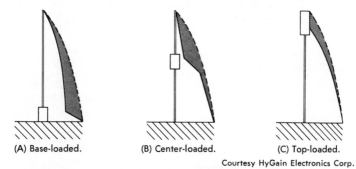

SOLID LINE: CURRENT DISTRIBUTION IN A SHORT ANTENNA
DOTTED LINE: CURRENT DISTRIBUTION IN A FULL LENGTH ANTENNA

(A) Base-loaded.   (B) Center-loaded.   (C) Top-loaded.

Courtesy HyGain Electronics Corp.

Fig. 3-9. The effect of a loading coil on current distribution in an antenna.

However, you cannot obtain something for nothing; so, increasing the field strength at one point is made possible only at the expense of field strength elsewhere. This can be

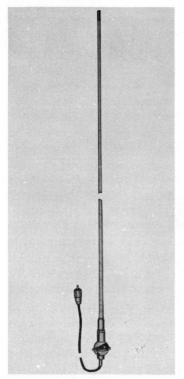

Fig. 3-10. A continuously loaded mobile whip (Model M-52).

Courtesy The Antenna Specialists Co.

seen from Fig. 3-11. Although somewhat exaggerated for illustration, these radiation patterns show the distribution of radio energy from two different antennas—one omnidirectional and the other unidirectional—supplied with the same amount of power. The antenna on the left (omnidirectional) provides no gain. Its radiation pattern extends an equal distance from the antenna in all directions. The unidirectional antenna, however, is designed essentially to squeeze this pattern together in front of the antenna, thereby making it elongated as shown. This field pattern covers the same amount of overall area as the one on the left, except that here almost all the radiated energy is "focused" in one direction. Therefore, rather than providing equal field strength in all directions, the unidirectional antenna takes field strength from places where it is not needed and directs it to another area where it is.

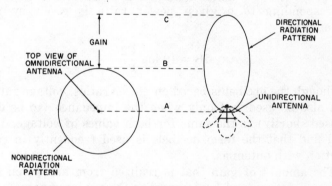

Fig. 3-11. Antenna radiation patterns showing how gain is achieved.

Notice that both antennas are located directly opposite each other as indicated by dotted line A. The distance between dotted lines B and C represents the forward gain of the directional antenna over the one at the left which has no gain. Obviously, to realize this gain a unidirectional antenna must be pointed toward the station with which communication is desired. This does not mean the antenna will not radiate or receive signals off the side or back. It will, but not as efficiently as in the forward direction. In reality there may be one or more "lobes" off the back side of the antenna pattern, as shown by the dotted lines. These responses do permit short-distance communications without the necessity of swinging the antenna around; but for optimum performance, the transmitting or receiving station must be in the path of the frontal lobe.

The term "front-to-back ratio," in connection with unidirectional antennas, is the ratio of power radiated in the direction of maximum response to that radiated in the opposite direction. The same ratio is used with respect to the receiving ability of an antenna. Here, it is a measure of the reduction in received signal when the antenna is changed from the direction where maximum response is obtained to the opposite direction. The front-to-back ratio of an antenna is expressed in decibels.

The Decibel

Decibels are units for the logarithmic expressions of power ratios. The number of decibels (abbreviated dB) corresponding to a given power ratio is given by the formula:

$$dB = 10 \log \frac{P2}{P1} \quad \frac{output\ power}{input\ power}$$

Although the decibel is based on power ratios, voltage ratios can also be used, but only when the impedance (to be discussed shortly) is the same for both values of voltage. You will find that the term decibels is used frequently in connection with antennas.

The amount of gain that is realized from any given antenna is determined by comparing the performance of the antenna in question with that of a standard antenna and then by expressing this figure as a ratio of the power levels required to produce equivalent field strengths. The gain, then, in decibels equals ten times the log of this power ratio. A decibel chart appears in Table 3-1 for your convenience in figuring effective radiated power (ERP), etc.

Omnidirectional Gain

During the previous discussion for Fig. 3-11, we mentioned that this omnidirectional antenna had no gain. Before any gain can be realized, an antenna must be able to concentrate a portion of its radiated rf energy in a manner that adds to, or reinforces, the existing energy already fol-

Table 3-1. Decibel Chart

dB	Voltage Ratio		Power Ratio	
	Gain $\dfrac{E_1}{E_2}$	Loss $\dfrac{E_2}{E_1}$	Gain $\dfrac{P_1}{P_2}$	Loss $\dfrac{P_2}{P_1}$
1	1.12	0.891	1.26	.794
2	1.26	0.794	1.58	.631
3	1.41	0.708	1.99	.501
4	1.58	0.631	2.51	.398
5	1.78	0.562	3.16	.316
6	1.99	0.501	3.98	.251
7	2.24	0.447	5.62	.178
8	2.51	0.398	6.31	.158
9	2.82	0.355	7.94	.126
10	3.16	0.316	10.00	.100
11	3.55	0.282	12.60	.079
12	3.98	0.251	15.80	.063
13	4.47	0.224	19.90	.050
14	5.01	0.199	25.10	.040
15	5.62	0.178	31.60	.032
16	6.31	0.158	39.80	.025
17	7.08	0.141	50.10	.020
18	7.94	0.126	63.10	.016
19	8.91	0.112	79.40	.013
20	10.00	0.100	100.00	0.10

lowing the desired path. As you will recall from the discussion of radio waves in Chapter 2, the portion of the radiation that travels upward in the form of sky waves generally serves no useful purpose and is only dissipated in space. An omnidirectional antenna designed to lower the angle of radiation and hence divert the energy that would normally be lost as sky waves along ground-wave paths will, in effect, produce an omnidirectional gain. Just how much gain, of course, will depend on how much additional energy is directed along these paths. An antenna with a gain of 3 dB effectively doubles transmitter power. All things being equal, a class-D transmitter having a power input of 5 watts and feeding an antenna having a gain of 3 dB will radiate a signal that will appear to have originated from a transmitter having a 10-watt input and feeding an antenna with zero gain. Unfortunately, the communicating range does not increase in direct proportion to the transmitter power, although it will be extended.

Radiation Patterns

Radiation patterns frequently accompany antenna data and are very useful because they show at a glance the

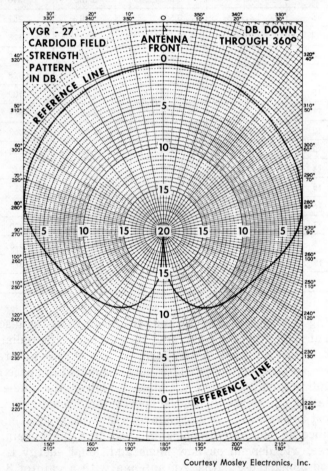

Fig. 3-12. Horizontal radiation pattern for the antenna in Fig. 3-13.

radiation and reception characteristics of an antenna. These patterns are obtained by plotting the results of field-strength measurements taken at various points throughout the area of coverage. A typical radiation pattern is shown in Fig. 3-12. This pattern, called a cardioid (meaning heart shaped), depicts the operating characteristics of the antenna shown in Fig. 3-13. In the latter figure, you can see the relationship between the antenna elements and the radiation pattern. This cardioid vertical antenna provides a radiation pattern with gain over an azimuth of 180°. The gain in the direction of maximum signal (indicated by the arrow) is 3 dB ±90° from the axis of the main forward lobe. The 3-dB gain doubles the effective radiated power in the forward direction. The front-to-back ratio for this antenna in this instance is 15 dB.

We might also point out that even though these are referred to as radiation patterns, for all practical purposes they apply equally well to the reception characteristics of an antenna. This means, then, that a high-gain antenna will receive radio signals more effectively than one providing little or no gain.

In reality the radiation pattern of a given antenna will not appear the same in every installation. This is due to a number of factors. At 27 MHz (the class-D frequencies), for example, the height of an antenna above the ground is a major factor in determining its radiation pattern. What is more, the electrical characteristics of the ground beneath the antenna also have considerable influence on the pattern. The height of an antenna can be such that ground reflections will reinforce the vertical radiation at the most desirable angle. In CB operation, the angle of radiation is generally kept as low as possible to direct most of the radiation along ground-wave paths. The physical height needed to obtain certain characteristics is primarily dependent on the frequency of the radio signals. An antenna designed for 27-MHz operation should be at least one wavelength, or approximately 36 feet above the ground. Unfortunately, this cannot always be attained in CB radio because of the antenna-height limitations imposed by the FCC. (These will be discussed in connection with base-station installations in Chapter 5.)

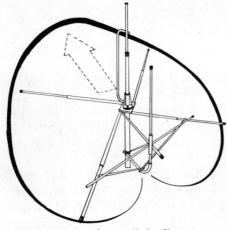

Fig. 3-13. Relationship between the elements of the VGR-27 antenna and the Cardioid radiation pattern.

Courtesy Mosley Electronics, Inc.

Parasitic Elements

Parasitic elements enable both vertical and horizontal antennas to provide gain. Moreover, these elements can make a vertical antenna directiònal, as seen in Fig. 3-13, and an already directional horizontal antenna even more directional. Parasitic elements are passive; that is, unlike the

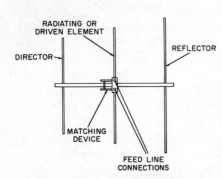

Fig. 3-14. Relationship between the elements of a 3-element beam.

driven element, they are not connected to the feed line but instead receive power by induction or radiation from the driven element. This power is radiated from the parasitic element with the proper phase relationship to achieve the desired effect. A parasitic element that reinforces radiation on a line pointing to it from the antenna is called a director. When the opposite is the case, the parasitic element is

48

termed a reflector. Just which term will apply depends on the tuning of the parasitic element.

Fig. 3-14 shows the relationship between the driven and the parasitic elements of a three-element beam antenna. Generally, only one reflector is used with this type of antenna, although there may be several directors. The various elements can usually be identified by their lengths. Directors, for example, are shorter than the driven element, while the reflector is longer than the driven element. Moreover, the position of the directors indicates the front of the antenna.

Fig. 3-15 shows how adding directors to an antenna affects the radiation pattern. By itself, the driven element (a dipole) displays a figure-8 pattern (Fig. 3-15A). Adding a reflector and two directors, as in Fig. 3-15B, produces forward gain and at the same time reduces the field strength off the back and sides of the antenna. As additional directors are added (Fig. 3-15C and D), the pattern becomes more elongated as shown. This, of course, represents more forward gain; however, notice that as the pattern becomes longer, field strength is reduced along the sides and off the back of the antenna. The frequency for which an antenna is designed partially determines the number of elements that can be used. At uhf frequencies it is not uncommon to find yagis with 8 to 10 elements or more. However, the spacing required between the elements at the much lower class-D frequencies limits their number.

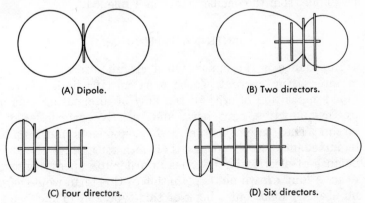

(A) Dipole. (B) Two directors.

(C) Four directors. (D) Six directors.

Fig. 3-15. The effect on the radiation pattern when directors are added to an antenna.

The gain of an antenna with parasitic elements varies not only with the spacing of these elements but also with their tuning. Therefore, at any given spacing there will be one tuning condition which will provide maximum antenna gain. Seldom if ever does the maximum front-to-back ratio occur under the same conditions that permit maximum forward gain. It follows, then, that the gain, spacing of elements, and front-to-back ratio of any given antenna necessitate that a number of compromises have to be considered in the antenna design.

Effective Radiated Power

There is a difference between the output power of a transmitter and the effective radiated power. The latter, as its name implies, is the effective radiation from the antenna itself. When a gain antenna is employed, the effective radiated power can be considerably higher than the actual transmitter output power. In fact, it is possible to multiply power by 8 to 10 times or even more, depending on the type of antenna, installation, etc. Suppose that a CB transmitter has 3 watts of rf power output. If this 3-watt output were fed to an antenna with a 3-dB gain (and assuming there were no losses, of course), the effective radiated power would be doubled. If this same 3 watts were fed (again without losses) to a different antenna having a gain of 7 dB, the effective radiated power would be multiplied by five. The effective radiated power for various gains can be found from the decibel chart in Table 3-1.

ANTENNA IMPEDANCE

Impedance is the opposition a circuit offers to the flow of alternating current. To be more specific, it is the combined opposition offered to the flow of alternating current by the inductive, capacitive, and resistive properties of a circuit. The statement that two components must match in impedance simply means that both must have the same balance of electrical properties in order to appear the same to ac. A long drawn-out explanation of this term would serve no useful purpose in a book of this type and in fact would only tend to cloud the issue. Therefore, for simplicity let us simply say that, in general, impedance is to ac what resistance is to dc.

Impedance is particularly important in connection with the transfer of rf power to the antenna. Every antenna, regardless of type, will have a certain value of impedance. It may be somewhat different at various points, but it does have considerable bearing on the amount of power that will be transferred from the transmitter to the antenna. The impedance of an antenna at any given point is the ratio of the voltage to the current at that point. It may not be the same at all points, either. For example, an antenna may have a certain impedance at the point where the feed line (or transmission line, as we will refer to it) connects and another quite different value near the tip. Antenna impedance is important in both the transmission and the reception of radio signals, but it is especially important in the former sense. Before radio energy from a transmitter can be effectively radiated into space, it must be fed to some type of load. The output circuit of the transmitter, the transmission line, and the antenna itself, all have a certain value of impedance; and only when these impedances match will maximum rf energy be transferred between the transmitter and the load, or antenna. When the transmission line is connected between the antenna and the receiver, the receiver input circuit—not the antenna—is the load because now the power is transferred from the antenna to the receiver.

Standing Waves

When the impedances along a given path are equal, the ratio of voltage to current at any point along this path will likewise be the same. However, should this path be terminated with a device having a different impedance, some of the current will reverse direction when it encounters this dissimilar impedance. A transmission line is an electrical path terminated at one end by the transmitter output circuitry and at the other end by the antenna. Now, if the impedance of the antenna system is different from that of the transmission line, a portion of the rf energy flowing to the antenna will be reflected back toward its source because of this mismatch. Just how much energy is reflected will depend, of course, on how greatly the impedance differs. This reflected energy results in standing waves along the transmission line. This characteristic, commonly expressed as the standing-wave ratio (or simply as swr), is the ratio of maximum to minimum current along the line. The same ratio holds true for voltage as well. For maximum effective radiated power, the swr must be kept as low as possible, particularly in CB radio, because there is very little rf power available to begin with. If the swr is to be kept to a minimum, the load (antenna) at the point of connection to the transmission line must appear purely resistive. If the load is purely resistive and equal in value to the characteristic impedance of the transmission line, there will be no standing waves. This ideal condition is generally difficult to obtain because of a number of factors which affect impedance, such as the type of antenna, its tuning, its height above ground, and the transmission line employed.

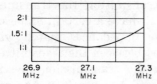

Fig. 3-16. A typical swr curve.

Communications antennas for CB as well as commercial operation are generally designed to operate over an entire band of frequencies. In order to tune such an antenna so that it responds efficiently at any given frequency (or over several frequencies), certain elements are made adjustable. If their lengths are incorrect, a mismatch will occur, resulting in standing waves and hence lost power. When operation is required over several frequencies without readjusting for each, the antenna is generally tuned to provide minimum swr at the center frequency of the group. This permits compromise operation over all channels, but the antenna will be most efficient only at the exact frequency for which it is tuned. The lower the swr ratio, the more efficient the antenna system. An example of an swr curve for a typical 27-MHz antenna is shown in Fig. 3-16. As you can see, optimum efficiency is obtained at 27.1 MHz (the frequency for which the antenna is tuned), and losses increase on either side of this frequency. An swr of 1:1 is considered ideal. Additional information on the cause, reduction, and measurement of standing waves is covered in a later chapter.

Impedance-Matching Devices

Practically all CB equipment is designed to operate into a transmission line and an antenna having a 50- to 52-ohm impedance. Should either have a characteristic impedance of a different value, some form of impedance-matching device will have to be used. As the various antennas are discussed in the following chapters, a number of these devices will be encountered.

CHAPTER 4

Typical CB Antennas

Practically all CB antennas can be classed as either whip, coaxial, dipole, ground plane, or beam. There are, of course, a number of variations within these groups, and some of these antennas are directly interrelated. For example, a beam antenna is nothing more than a simple dipole (two-piece driven element) used in conjunction with several parasitic elements. A further look will reveal that a ground-plane antenna is essentially a simple vertical antenna with radials extending outward from its base. So you can see there is a definite relationship between the various classes.

Some antennas consist of only a single element, while others employ several. What is more, antennas can be designed for either horizontal or vertical polarization, or both. They may be of a fixed length or adjustable, and they may or may not employ loading coils. In addition, antennas may be constructed from one or more materials such as steel, brass, aluminum, or fiber glass, and they are available with almost any type of mounting for any kind of installation.

Now let us look at some typical CB antennas within these different classes to see how they compare with each other physically, electrically, and in performance.

THE VERTICAL WHIP ANTENNA

The whip is nothing more than a single vertical rod (usually rather thin) composed of a suitable conducting material and cut to a specific length. This radiator is classified as a vertical polarized antenna, and the rf signal is fed to the bottom end. Whips are generally constructed of stainless steel or some other strong yet flexible material. The term whip was adopted because of the way this element flexes, sways, and whips about when subjected to abrupt movements. In recent years a number of whips have been constructed of fiber glass which is somewhat less flexible. Such antennas usually have a larger diameter than stainless-steel whips, and since fiber glass is not a conductor of electricity, they will either have a conductor running through the center or will be wound in a spiral around the outer surface.

The vertical whip may be almost any length, depending on the frequency at which it is designed to operate. At uhf frequencies, for example, a whip that is cut to a full wavelength would probably not present any length problems, whereas a half-wave whip at 27 MHz might. The majority of whips for class-D operation are cut to a quarter-wavelength, and some are cut even shorter.

Quarter-wave whip antennas are used extensively in all classes of CB operation because they have a number of desirable features. For one thing they are well suited for installations where space is a problem. Also, their omnidirectional response means that signals can be radiated and received equally well in all directions. Unlike unidirectional antennas, the whip does not have to be rotated to any certain direction before efficient operation can be achieved. Moreover, through proper design, the angle of radiation can be lowered to the point where an omnidirectional power gain can be realized. Vertical whip antennas are used in practically all mobile and portable operations.

A quarter-wave antenna can be fed directly with a 50-ohm coaxial transmission line or fed with one of a different impedance, provided it is used with a suitable impedance-matching network. The 27-MHz whip antenna in Fig. 4-1 is mounted directly on the transceiver by a swivel joint, is telescopic, and has maximum and minimum lengths of 45 and 15 inches. A base-loading coil, enclosed in polypropylene plastic, is employed to provide more efficient operation with the shortened whip.

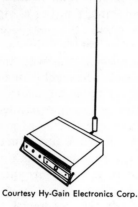

Fig. 4-1. A typical base-loaded telescopic whip.

Fig. 4-2. A base-loaded mobile whip antenna (Model M-125).

Fig. 4-3. Model HW-11 Heliwhip mobile antenna.

The whip shown in Fig. 4-2 is also shorter than a quarter-wavelength and employs base loading. Unlike the previous example, however, it is designed for mobile operation and has a fixed length of 46 inches. This particular antenna is designed for installation on the roof of a vehicle.

A great many mobile antennas, especially the loaded types, are made of fiber glass. As an example, the 27-MHz mobile *Heliwhip* shown in Fig. 4-3 is designed to replace

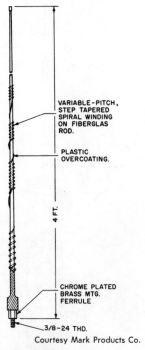

VARIABLE-PITCH, STEP TAPERED SPIRAL WINDING ON FIBERGLAS ROD.

PLASTIC OVERCOATING.

4 FT.

CHROME PLATED BRASS MTG. FERRULE

3/8-24 THD.

Courtesy Mark Products Co.

Fig. 4-4. Detailed view of the loading method used with the Heliwhip.

Courtesy Mosley Electronics, Inc.

Fig. 4-5. A base-loaded telescopic whip with suction-cup mounting.

the conventional stainless-steel quarter-wavelength (108-inch) whips, as well as the base-loaded types. It is constructed of molded fiber glass, has an overall length of 4 feet, and is continuously loaded by the spiraled conductor shown. (A more detailed view of this antenna can be seen in Fig. 4-4.) Fig. 4-5 shows still another example of a class-D whip in the "less-than-a-quarter-wavelength" category. This base-loaded antenna is telescopic and has a maximum length of 57 inches. A large suction cup at the base permits it to be mounted on top of the transceiver, on a desk or any other smooth surface.

COAXIAL

At high frequencies, it is desirable to keep the angle of radiation as low as possible since the sky waves generally

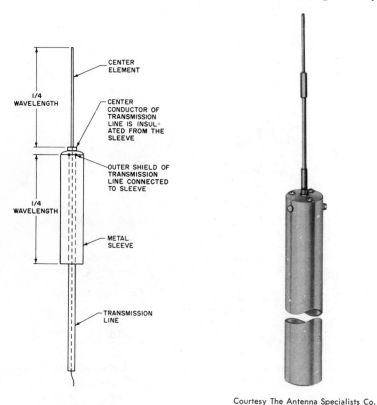

Fig. 4-6. Basic construction of a coaxial antenna.

Courtesy The Antenna Specialists Co.

Fig. 4-7. Typical coaxial base-station antenna.

dissipated at these frequencies serve no useful purpose. The most practical solution is to divert this otherwise wasted skywave energy along paths where it can be put to good use. The coaxial antenna (often referred to as a "thunderstick") has a good low-angle radiation pattern and is therefore quite suitable for CB communications. A great many coaxial antennas are currently being used in base, or fixed-station, installations.

The basic construction of a coaxial antenna is shown in Fig. 4-6. Like practically all other CB radio antennas, this antenna is fed by a concentric, or coaxial, transmission line comprised of a center conductor surrounded by a dielectric material, which, in turn, consists of a shielding braid and a vinyl jacket. Essentially, the transmission line is passed upward, through a metal sleeve, to an element at the top. The center conductor of the coaxial line is connected to this element, thus effectively extending this conductor one-quarter wavelength beyond the end of the transmission line. The outer shielding braid of the transmission line is connected to the top of the metal sleeve as shown. Also, notice that the center element and the sleeve are insulated from each other. This, then, represents a half-wave antenna since both the extended element and the metal sleeve are a quarter-wavelength long and in opposite directions from the feed point.

The sleeve acts as a shield around the transmission line; as a result, very little current is induced on the outside of this line by the antenna field. Coaxial antennas used in CB radio are generally designed to present a 50- to 52-ohm impedance to the line. The line itself is nonresonant because its characteristic impedance is the same as the center impedance of the antenna.

Fig. 4-7 shows a typical 27-MHz coaxial antenna designed primarily for base-station installations. Antennas of this type require very little space, are easy to mount, and provide a low-angle radiation pattern. This half-wave antenna has an overall length of 18 feet, and it can be mounted on a ¾-inch iron pipe. The 9-foot vertical radiator is jointed and made of solid aluminum; the skirt, also of aluminum, is 9 feet in length and 2 inches in diameter.

The coaxial antenna in Fig. 4-8 is for class-D operation; however, a similar unit can be obtained for use with class-A CB equipment. The antenna shown here provides an efficient low angle of radiation and can be fed with a 52-ohm transmission line. Being cut to the center of the class-D band, it permits compromise operation over all class-D frequencies.

A vswr of 1.1:1 can be achieved at the resonant frequency. This antenna is constructed of tempered aluminum and weighs only 5 pounds.

As mentioned previously, there are quite a few modifications of the basic antenna designs. A good example is the "stacked' coaxial antenna shown in Fig. 4-9. By means of the

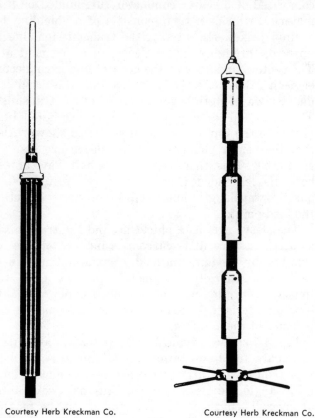

Courtesy Herb Kreckman Co.

Fig. 4-8. Model CO-CBA coaxial antenna for 27-MHz operation.

Courtesy Herb Kreckman Co.

Fig. 4-9. A stacked coaxial antenna which provides a 6-dB omnidirectional gain at the class-A CB frequencies.

additional skirts and the horizontal ground-plane radials at its base, this antenna provides such a low angle of radiation that a 6-dB omnidirectional gain is realized. This unit is designed for uhf communications at the class-A CB frequencies, is of all-brass construction, and weighs 10 pounds.

Fig. 4-10. The basic dipole antenna.

DIPOLE ANTENNAS

A dipole antenna is nothing more than a two-piece driven element fed at the center, as shown in Fig. 4-10. It may be a half- or a quarter-wavelength, or—like the vertical whip—it can be shorter physically than a quarter-wavelength and employ loading coils, as shown in Fig. 4-11A. This type of dipole, commonly referred to as "rabbit ears," is used primarily for portable or indoor operation and is limited to short-range communications. The telescoping whips shown here extend from 15 to 45 inches. The loading coils include an L-matching network that provides a 50-ohm impedance match to the line. A suction-cup base permits mounting on a wall, a table, or the top of the transceiver itself. Moreover, in apartment buildings and other locations where indoor space is limited and most outdoor installations are prohibited, the antenna can be fastened to a windowpane, as shown in Fig. 4-11B.

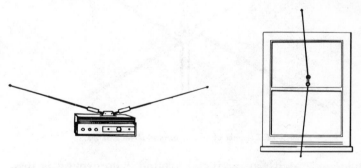

(A) Mounted on top of the transceiver.　　(B) Mounted on a windowpane.

Courtesy Hy-Gain Electronics Corp.

Fig. 4-11. Telescopic dipole antenna with loading coils and suction-cup mounting.

Portable and indoor dipole antennas appear in several forms; however, most will be similar to the example shown. Its radiation pattern is similar to a figure eight, and the transmitted waves comprise components of both horizontal and vertical polarization. Most indoor dipoles for class-D equipment are engineered to match, or at least present as nearly as possible, a load impedance of 50 to 52 ohms.

GROUND PLANE

Of the five basic antenna groups listed at the beginning of this chapter, the ground plane is the most widely used for class-D base (fixed) station operation. One reason is that its angle of radiation can be lowered to the point where the radiated power can effectively be doubled, or increased even more, without destroying its omnidirectional characteristics. Basically, the ground-plane antenna is a vertical radiating

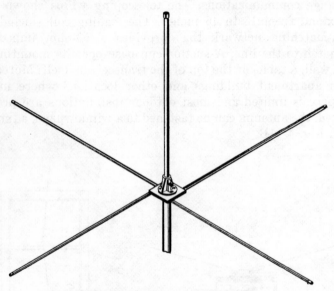

Fig. 4-12. Example of a basic ground-plane antenna.

element used with an artificial ground. The ground is provided by three or four radials extending outward at right angles from the base of the driven element, but isolated from it (Fig. 4-12).

A simple (single-element) vertical quarter-wave antenna must have a good physical ground before it will provide the desired low angle of radiation. Moreover, its height has a considerable effect on the pattern. By the addition of ground-plane radials, however, it can be made to provide low-angle radiation regardless of its actual height above the ground. However, to obtain the true ground-plane effect, the radials should be a quarter wavelength or more above ground. The radials may be at right angles to the driven element or less, depending on the design of the antenna and on the radiation pattern desired. Ground-plane antennas provide exceptional communicating range, especially below 30 MHz. This does not necessarily limit these antennas to class-D frequencies alone; some are designed for CB communications in the uhf band.

Courtesy Hy-Gain Electronics Corp.

Fig. 4-13. The "Penetrator" colinear antenna provides 5.1 dB
of omnidirectional gain at the class-D frequencies.

The ground-plane antennas currently finding widespread usage in CB radio are anywhere from a quarter to nearly a full wavelength. Generally, they are of hollow aluminum tubing and large enough to provide sufficient strength yet small enough to present minimum wind resistance. (A compromise is often made in this respect.) The driven element and radials are almost always self-supporting. The radiation resistance of the ground-plane antenna varies somewhat with the diameter of the driven element, and a matching device is sometimes needed in order to present a 50- to 52-ohm load impedance to the transmission line.

Fig. 4-13 is an example of a popular colinear ground-plane antenna designed for class-D CB operation. It con-

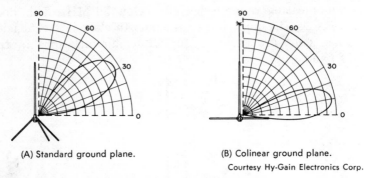

(A) Standard ground plane. (B) Colinear ground plane.

Courtesy Hy-Gain Electronics Corp.

Fig. 4-14. Typical vertical radiation patterns.

sists of a ⅝-wavelength driven element over a ground plane which provides a second ⅝-wavelength mirror image. Together these develop a broadside omnidirectional gain of 5.1 dB in field intensity. Moreover, this gain is achieved at a lower vertical angle of radiation than would normally be obtained with a standard quarter-wave ground-plane antenna. Fig. 4-14 shows a comparison between the vertical-radiation angles of a standard ground plane (A) and the colinear (B). Since the reciprocal characteristics of a communication antenna make it respond equally well during transmission or reception, this antenna is capable of extending the reliable range of CB communications. The overall height of its radiator is 22 feet 9½ inches, the ground-plane radials are 9 feet long, and all elements are of high-strength aircraft-type aluminum tubing.

Fig. 4-15 illustrates another popular CB ground-plane antenna known as the "Mighty Magnum III." This antenna has an overall height of 18 feet and employs shortened radials measuring only 60 inches in length. This antenna has a very low angle of radiation and provides 4 dB of omnidirectional gain. Later in this chapter some interesting variations in basic antenna designs will be shown.

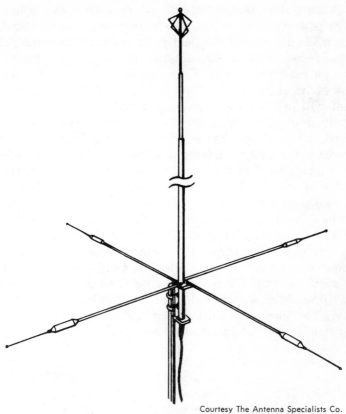

Fig. 4-15. "Mighty Magnum III" features shortened power-tip radials, low angle of radiation, and 4 dB of omnidirectional gain.

BEAM ANTENNAS

A beam antenna is generally composed of a series of elements arranged to provide a high power gain. It usually employs a dipole for a driven element and is used in conjunction with several elements known as parasitics. By combining these elements into a suitable array and properly spacing them, it is possible to make their radiation add up along a single direction and form a beam (hence the term, "beam" antenna). In this way, a power gain is obtained in one direction, but only at the expense of radiation in other directions. (Parasitic elements, radiation patterns, and antenna gain were discussed in Chapter 3.)

Beam antennas are capable of providing exceptionally high power gains and can also be very directional. Usually, the more directional the field pattern, the higher the gain. Also, remember that the radiation pattern of an antenna is not only an indication of its transmitting characteristics but of its receiving qualities as well. In CB radio, beams that will provide a power gain of 9 dB (which will effectively multiply transmitter power approximately eight times) are not uncommon. If we use this figure as an example, a transmitter having 4 watts of output and feeding such an antenna would, in effect, perform like one with 32 watts of output. An antenna capable of such performance is shown in Fig. 4-16. As you can see, this 27-MHz three-element beam may be mounted either horizontally (Fig. 4-16A) or vertically (Fig. 4-16B). Proper operation depends on (1) element spacing, (2) element length, and (3) impedance matching. Here an 8-foot boom is used for optimum element spacing. When the spacing is correct, the antenna will provide maximum gain and front-to-back ratio. Moreover, the elements must be of the correct length (not shortened) for maximum efficiency.

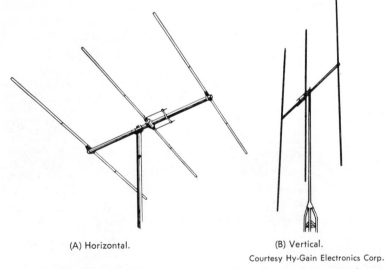

(A) Horizontal.   (B) Vertical.

Courtesy Hy-Gain Electronics Corp.

Fig. 4-16. This 3-element beam (Model 113-B) is designed for class-D operation, has a forward gain of 9 dB, and may be mounted either horizontally or vertically.

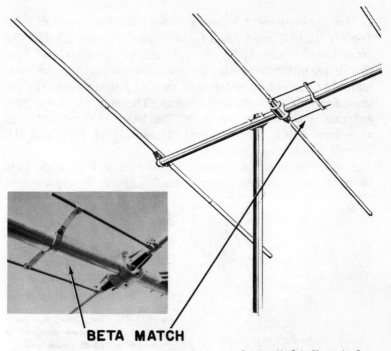

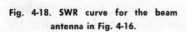

BETA MATCH

Courtesy Hy-Gain Electronics Corp.

Fig. 4-17. The Beta match employed on the antenna in Fig. 4-16.

When a beam antenna is tuned for maximum gain, the radiation resistance at the feed points is roughly 20 ohms. Hence, before optimum performance can be achieved, some type of impedance-matching device is generally necessary to permit the use of a 52-ohm transmission line. The matching system used with this antenna develops 52 ohms, thereby providing the proper match for RG-58/U or RG-8/U coaxial cable. This matching device (known as the *Beta* match) is shown in Fig. 4-17. The driven element is center-grounded to the boom through the matching system for lightning protection; yet this ground path is effectively open to rf.

Fig. 4-18. SWR curve for the beam antenna in Fig. 4-16.

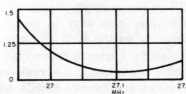

This three-element beam provides a forward gain of 9 dB (power times 8) and has a front-to-back ratio of 25 dB. The front-to-side ratio is 50 dB. Adjustable elements and *beta* match permit tuning the antenna and adjusting for final matching, to obtain minimum swr and to compensate for the effects of surrounding objects. The swr curve for this antenna is shown in Fig. 4-18. The beam is constructed of aluminum tubing; the longest element is 18 feet, and the overall weight is 15 pounds.

A much larger beam antenna is shown in Fig. 4-19. This beam, known as the "Long John," employs 5 wide-spaced

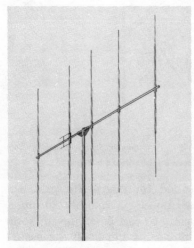

Fig. 4-19. The "Long John" is a wide-spaced, 5-element beam designed to provide a 12.3-dB power gain.

Courtesy Hy-Gain Electronics Corp.

elements mounted on a 24-foot boom and provides a power gain of 12.3 dB. As mentioned in a previous chapter, even more gain can be achieved by stacking a pair of beams. Fig. 4-20 shows a typical installation using a stacked array of vertical beams.

Beam antennas are becoming more and more popular in CB radio, and when you consider some of their desirable features, it is easy to understand why. Much, of course, depends on the type of operation desired. Beams are best suited for communications between fixed stations (point-to-point operation), whereas omnidirectional antennas are favored for base-to-mobile and mobile-to-base operations.

The radio signals from an omnidirectional antenna, for example, will be received equally well by stations in other directions. However, at the same time a beam antenna is providing a forward gain in one direction, it is radiating much less off its back and sides. Therefore, a beam antenna will cause less interference with surrounding CB stations operating on the same frequency than an omnidirectional antenna will. Sometimes the front-to-back or front-to-side ratios of a beam are so high that it is difficult to work even the "locals" unless the beam is pointed toward them.

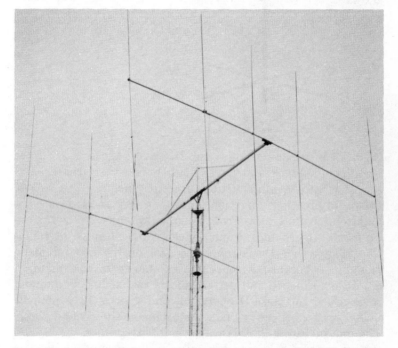

Fig. 4-20. Typical installation using a stacked array of wide-spaced, 5-element vertical beams.

VARIATIONS IN THE BASIC DESIGNS

This discussion will be devoted to some of the interesting variations of the basic antenna designs that have been covered. So many different antennas are being produced for CB radio that it is virtually impossible to describe

Fig. 4-21. Known as the Co-Plane, this combination coaxial/ground-plane antenna (Model CP-CBA) provides a 3-dB omnidirectional gain at 27 MHz.

Courtesy Herb Kreckman Co.

them all in a book of this size. However, those which have been included here and throughout this book are representative of what is available at the present time.

One interesting variation, shown in Fig. 4-21, is a combination coaxial and ground-plane antenna known as a *Co-Plane*. It provides a much lower vertical angle of radiation than either the straight coaxial or the ground plane alone; and because it concentrates energy in this manner, a 3-dB omnidirectional power gain is realized. This antenna is designed for class-D operation and weighs 11½ pounds.

In Fig. 4-22 there are two variations of the ground plane. The one in Fig. 4-22A is called a folded ground plane, and the antenna in Fig. 4-22B is called a duo ground plane. The advantages these antennas have over the basic designs are those of improved radiation pattern and lower vertical angle of radiation. An even-lower angle of radiation is achieved by the antenna shown in Fig. 4-23. This antenna provides an omnidirectional gain of 5 dB over a basic ground plane.

In recent years, a number of base-station antennas have been developed to provide high gain and a selection of either horizontal or vertical polarization. Some are even designed

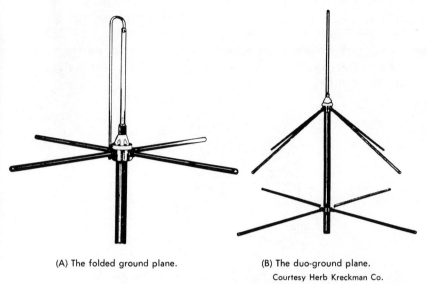

(A) The folded ground plane.

(B) The duo-ground plane.

Courtesy Herb Kreckman Co.

Fig. 4-22. Two more variations of the ground-plane antenna.

to radiate both of these simultaneously. Fig. 4-24 illustrates some of the most recent bipolarization designs for class-D operation.

ANTENNA SELECTION

The primary objective of any two-way radio installation is a highly efficient, yet low-cost, reliable system of communications that will make good use of the airways. And, of course, the antenna chosen for the installation will have a considerable influence on the effectiveness of the system. One of the biggest problems facing CB radio today is limited range—which in many instances can be directly attributed to improper selection and installation of the antenna.

Base-Station Antennas

A number of factors must be considered in selecting an antenna, although the final selection depends on each individual's needs. One of the major considerations is whether the intended operation is fixed, mobile, or portable. Of course, there are a great many variations among the antennas designed for a specific class of operation. The type of two-way radio equipment with which the antenna is to be used is also a contributing factor to the selection. Moreover, each fixed station presents a different set of requirements and therefore can be considered a custom antenna installation.

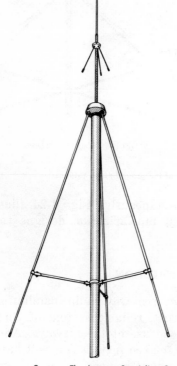

Fig. 4-23. This antenna, the M-400 "Starduster," provides an omnidirectional gain of 5 dB over a basic ground-plane antenna.

Courtesy The Antenna Specialists Co.

When new CB radio equipment is purchased, the antenna may be supplied or it may be available as an accessory. In some instances, the supplied antenna is actually a physical part of the equipment itself, as in the case of most pocket-sized transceivers. If portable operation is intended and only close-range communication is required (a mile or so up the street or even in the same building), such operation would hardly call for a complex rotary-beam antenna mounted at the maximum legal height; one of the short

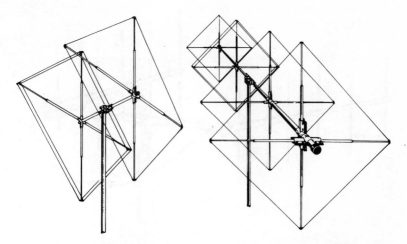

(A) The "Eliminator II" quad provides a power gain of 9 dB.

(B) "Big Gun II" employs quad construction with twin-driven, double-loop elements and provides a power gain of 14.5 dB.

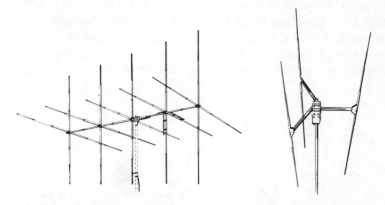

(C) The M-216 "Big Daddy" features yagi design and high gain and allows cross polarization.

(D) The "Super Scanner" Model MS 119 provides an omnidirectional gain of 5.75 dB and a directional gain of 8.75 dB.

Fig. 4-24. Examples of CB antennas that provide selection of either vertical or horizontal polarization.

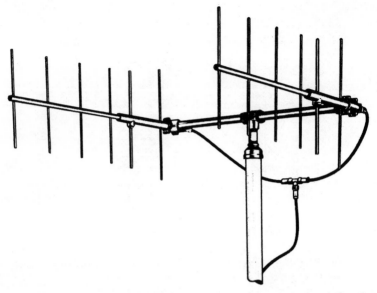

Courtesy Andrew Corp.

Fig. 4-25. This type 117 Yagi is designed for operation in the 450- to 470-MHz uhf band and has a forward gain of 12.2 dB.

portable whips that fasten directly to the transceiver will suffice. On the other hand, if long-range communications are desired but your CB station is set up in an apartment building where hole drilling for antenna installation is not permitted, one of the longer whips designed for temporary

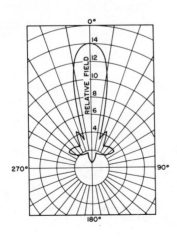

Fig. 4-26. Horizontal radiation pattern for the antenna in Fig. 4-25.

mounting (without drilling) on the outside window sill may be necessary. If you are fortunate enough to live on one of the upper floors, you may still be able to achieve more range than many of the ground stations because of the increased elevation of your antenna.

When maximum communicating range is desired from a fixed station, there are many things to consider. First of all, what type of coverage is required? If communications are to be with mobile units, some type of omnidirectional antenna should be employed. What is more, the distance these mobile units will be operating from the fixed station will determine whether or not an antenna that provides omnidirectional gain is necessary. If finances permit, such an antenna would certainly be desirable. Another important consideration is the amount of space available in which to make a permanent antenna installation. A ground plane, for example, is ideal for base-station installations, and many are capable of providing a high power gain. Where trees, buildings, and other objects will physically obstruct the horizontal ground-plane elements, however, a good coaxial antenna may be more advisable. This type of antenna is capable of providing the same omnidirectional pattern and gain, yet it requires less space than a ground plane. The location of a fixed-station antenna is also a contributing factor, as you can see. Quite often there is not much choice; you must make the best of what you have.

If communications over a considerable distance are intended between fixed points, a directional antenna can be used to great advantage. Not only does it provide substantial signal gain, but it generally discriminates against unwanted signals from other directions as well.

This is exemplified by the antenna shown in Fig. 4-25 and the associated radiation pattern in Fig. 4-26. This antenna system consists of a stacked 6-element yagi designed to operate at the class-A frequencies in the uhf region of the radio-frequency spectrum. From the radiation pattern, it is obvious that signals in directions other than that in which the antenna is pointed will cause little or no interference to other stations. Antenna polarization must also be considered. Practically all mobile units employ vertical antennas, whereas base stations may employ either a vertical or a horizontal antenna. So, if operation is intended between base stations only, by all means use a horizontally polarized antenna. Conversely, if mobile units are the primary concern, the antenna should be vertical. Communication, even though not as good, may still be satisfactory with local stations, even when cross polarization is used. Another alternative is to use one of the antennas that provides for selection of the desired polarization.

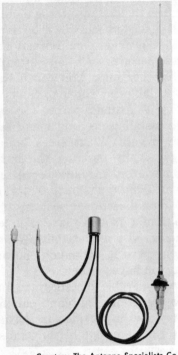

Fig. 4-27. An antenna for both CB and a-m operation. (Model M103).

Courtesy The Antenna Specialists Co.

If the base-antenna is directional, a rotor may be needed to swing the antenna around to the desired direction. This, of course, means an additional expense. Sometimes a used tv rotor can be obtained at a reasonable price. If expense is the determining factor, a ground plane may be more practical; it can provide a fair amount of gain but requires no rotor because of its omnidirectional pattern. The directional antenna, of course, does not interfere with surrounding stations as much.

Mobile Antennas

Some of the considerations just discussed apply also to mobile-antenna selection, but there are additional considerations. Proper selection of a mobile antenna is just as important as selection of the base-station antenna.

One of the primary factors is the location of the antenna on the vehicle. The metal surface of the vehicle itself acts as a ground plane for the quarter-wave whip antenna used in this type of operation. Where this antenna is mounted on the vehicle affects the ground plane and hence the radiation pattern. This will be discussed at length in Chapter 6. For now, let us say that the roof mount is best but is usually not practical in class-D operation because of the physical length of a quarter-wave whip. Shortened antennas (those using loading coils) are often more practical where appearance is the primary concern or where overhead obstructions may be encountered. Moreover, where too many antennas on a vehicle might be objectionable, a single antenna can be used for both CB operation and standard a-m broadcast reception. An antenna system of this type is shown in Fig. 4-27. It consists of a top-loaded antenna (resembling a regular broadcast type) and a coupling device. Together these are all that are needed for both types of operation.

Where mobile operation is only temporary or where it is not desirable to drill holes into the vehicle itself, one of the clamp-on type antennas may be employed. These are available as full-length whips, which may be clamped onto the bumper, or as short, loaded whips, which may be mounted on the drain gutter above the car door, on the trunk ledge or the roof (by suction cups), or to an existing a-m broadcast antenna.

CHAPTER 5

Base-Station Antenna Installation

Every antenna installation, no matter how simple, involves a certain amount of planning. Moreover, each installation presents a unique set of requirements and may thus be considered a custom installation. Two of the primary requirements are the type of antenna and its mounting location.

HEIGHT LIMITATIONS

In selecting a mounting location, height is of utmost importance. CB antennas cannot always be mounted as high as desired because of the FCC limitations on antenna structures. (See Part 95 of the *FCC Rules and Regulations*.)

With class-C, and -D stations operated at fixed locations, the antenna must not exceed 20 feet above the natural or man-made structure on which it is mounted. This does not necessarily mean the antenna cannot be over 20 feet above the ground. In fact, it can be hundreds or even thousands of feet above the ground and still be only 20 feet above its mounting. Fig. 5-1 illustrates how the 20-foot limitation applies. Antenna *A* is fastened to a mast mounted on the ground. The overall height from the ground to the top of the antenna is 20 feet, the legal limit. Now consider antennas *B* and *C*, both mounted atop existing man-made structures. Both provide a higher *effective elevation* over antenna *A*, yet neither exceeds the legal 20-foot limit above the mounting. To the right of antenna *A* in Fig. 5-1 is a building atop a large hill. One CB antenna (*D*) is mounted 20 feet above the roof, and another (*E*) is on the ground below. Notice that every antenna in this illustration is only 20 feet above its mounting, yet some provide much more effective elevation; the actual height of the antenna above its

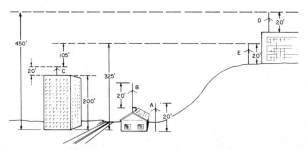

Fig. 5-1. Relationship between the 20-foot antenna-height limitation and the effective elevation.

mounting is not nearly as important as the effective elevation it affords.

One exception to the 20-foot limit is shown in Fig. 5-2. A CB antenna, when mounted on an existing antenna structure of another station, must not extend above this structure. Moreover, no such structure can be erected solely for mounting a CB antenna. If permitted, this would obviously defeat the purpose of restricting the antenna height.

The latest rule states that the highest point of an omnidirectional antenna or its supporting structure must not exceed 60 feet above the ground level and the highest point also must not exceed one foot in height above the established airport elevation for each 100 feet of horizontal distance from the nearest point of the nearest airport runway.

The height limitations on antenna structures for class-A stations are considerably more involved, but greater heights (up to 170 feet or more above ground level) are permitted. Antenna elevation is particularly important with class-A equipment because of the propagation characteristics at uhf. The higher transmitter power that is permitted class-A sta-

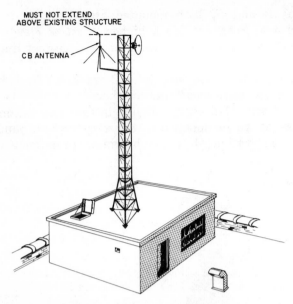

MUST NOT EXTEND
ABOVE EXISTING STRUCTURE

CB ANTENNA

Fig. 5-2. A CB antenna mounted in compliance with FCC regulations, on the existing antenna structure of another station.

tions cannot be fully utilized unless the antenna height permits a more or less line-of-sight path for the radio signals.

SELECTING THE MOUNTING LOCATION

After a suitable antenna has been selected, the next step is to decide where it should be mounted. Occasionally, low-hanging power lines, trees, and other obstructions may necessitate a compromise in the mounting location.

It is desirable, of course, to choose a spot that will afford maximum effective height, but this is not always possible. Perhaps the property on which the antenna is to be installed is rented or leased. In such cases an installation of this type is sometimes prohibited, especially where it requires the drilling of holes in a building. Here, the antenna may have to be mounted on the ground. While not the most desirable installation from the standpoint of elevation, it will at least satisfy the landlord. Then, too, the most advantageous location may already be cluttered up with television antennas. These can present a real problem in more ways than one.

Most tv antennas can be relocated without impairing the reception—in fact, it will often be improved. But convincing the antenna's owner that this is true is another matter.

Some tv antennas, especially those equipped with rotors, are quite top-heavy and therefore difficult to handle. To perform the job safely may require the services of several persons. Quite often the bolts and mounting straps are rusted so badly that they break and must be replaced at the expense of the person responsible for moving the antenna. Also, additional lead-in wire may have to be purchased. Therefore, unless such an installation is the most practical, it may be worthwhile to consider another mounting location for the CB antenna. Avoid fastening antennas, especially those that are self supporting, to old chimneys or even some of the newer ones for that matter. The stress presented by a strong wind may be sufficient to topple the chimney as well as the antenna. It would be better to mount the mast, or even a tower (provided it does not exceed the legal height limit) alongside the building and use the latter itself for support. Possibly there is another building nearby that will afford as much or even more antenna elevation, even though it may be farther from the radio equipment, or a friendly next-door neighbor may even permit you to mount the antenna on his property. As close together as many houses are in some of the larger cities, such a mounting may even bring the antenna closer to the radio equipment than would otherwise be possible. It is desirable that the antenna be mounted as near the radio equipment as possible in order to keep the transmission line short and thus cut down the losses it presents. At the same time, the mounting location should afford maximum effective antenna elevation. Quite often a compromise will have to be made between the two.

THE TRANSMISSION LINE

The electrical path between the antenna and the radio equipment is normally provided by a coaxial transmission line constructed as shown in Fig. 5-3. This cable is composed of either a solid or stranded center conductor and a braided outer conductor. The two are separated by some type of

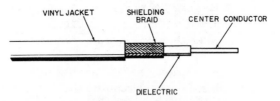

VINYL JACKET   SHIELDING   CENTER CONDUCTOR
               BRAID

DIELECTRIC

Fig. 5-3. Coaxial cable construction.

dielectric, and the whole thing is covered with a vinyl jacket. The outer conductor keeps out man-made electrical noise, which could otherwise produce considerable interference during reception. In addition, the shielding effect of the outer conductor tends to confine the rf energy within the cable and thus prevents it from being radiated before reaching the antenna. Unlike the coaxial cable shown in Fig. 5-3, some have more than one shielding conductor.

With all of its desirable features, the coaxial cable must still be kept as short as possible to minimize the rf losses that are incurred within it. Between any two conductors separated by a dielectric, there is always a certain capacitance, the amount depending on the size of the conductors, the spacing between them, and the dielectric constant of the material separating them. This capacitance results in a loss of rf energy. Some of today's dielectrics have been improved to the point where capacitance and hence rf losses are exceptionally low. The capacitive reactance (ac resistance) between these two conductors decreases as the frequency increases. In other words, more rf energy is lost at higher frequencies.

Almost all CB antennas are designed to accept a coaxial transmission line having a characteristic impedance of 50 to 52 ohms. Table 5-1 lists representative coaxial cables having an impedance within this range and gives the rf losses they present at various frequencies. These losses are usually expressed as so many decibels of attenuation per 100 feet. From the chart, you can see that not all cables use the same dielectric material; also, the amount of capacitance per foot varies according to the type of cable. Notice the difference between the losses of the RG-8/U and the RG-58A/U cables. At 10 MHz the RG-8/U presents a 0.55-dB attenuation per 100 feet of cable, as opposed to 1.6 dB for the RG-58A/U.

Table 5-1. Coaxial-Cable Characteristics

Type RG.../U	Impd. (ohms)	Capacity (μF per foot)	Diam. (inches)	Attenuation—dB per 100 ft					Remarks
				1 MHz	10 MHz	100 MHz	400 MHz	1000 MHz	
5A	50	29	.328	.16	.66	2.4	5.25	8.8	Small, low loss.
8	52	29.5	.405	.16	.55	2.0	4.5	8.5	General-purpose.
9	51	30	.420	.12	.47	1.9	4.4	8.5	General-purpose.
9A	51	30	.420	.16	.59	2.3	5.2	8.6	Stable attenuation.
14	52	29.5	.545	.10	.38	1.5	3.5	6.0	RF power.
16	52	29.5	.630	—	—	—	—	—	RF power.
17	52	29.5	.870	.06	.24	.95	2.4	4.4	RF power.
18	52	29.5	.945	.060	.24	.95	2.4	4.4	Polyethylene dielectric.
19	52	29.5	1.120	.04	.17	.68	1.28	3.5	Low-loss rf.
20	52	29.5	1.120	.04	.17	.68	1.82	3.5	Low-loss rf.
58A	50	28.5	.195	.42	1.6	6.2	14.0	24.0	Polyethylene dielectric.
87A	50	29.5	.425	.13	.52	2.0	4.4	7.6	Teflon dielectric.
117	50	29	.730	.05	.20	.85	2.0	3.6	Teflon and Fiber glass.
119	50	29	.470	—	—	—	—	—	Teflon and Fiber glass.
122	50	29.3	.160	.40	1.70	7.0	16.5	29.0	Miniature coaxial.
126	50	29	.290	3.20	9.0	25.0	47.0	72.0	Teflon and Fiber glass.
141	50	29	.195	.35	1.12	3.8	8.0	13.8	Teflon and Fiber glass.
142	50	29	.206	.35	1.12	3.8	8.0	13.8	Teflon and Fiber glass.
143	50	29	.325	.24	.77	2.5	5.3	9.0	Teflon and Fiber glass.
174	50	30	.10	—	—	—	19.0	—	Miniature coaxial.

The difference between these losses becomes even more significant as the class-A CB frequencies in the uhf range are approached. At 400 MHz, for example, the RG-8/U cable now has 4.5-dB attenuation, compared with a high of 14 dB for the RG-58A/U. The same cable presents a 24-dB attenuation per 100 feet at 1000 MHz.

This, then, should give you some idea of the amount of rf losses that can be incurred in a coaxial cable. These figures are not the same for all cables, even those with the same type number, because different manufacturers may use different dielectrics, etc.

ASSEMBLING AND TUNING ANTENNAS

After a desirable mounting location has been selected, the next step is to assemble (and in some cases tune) the antenna.

Assembling

Except for small, portable antennas and most mobile whips, practically all antennas are sold either collapsed or completely dismantled. Beams, for example, are usually intact but have their elements swiveled back parallel with the boom. Usually all that has to be done is to fold the elements out to their normal position (right angles to the boom) and secure them with the appropriate hardware (which is furnished). After all elements are properly positioned and tightened, the antenna can be fastened to the mast while the entire assembly is still on the ground. The transmission line is also connected to the antenna at this time. The method of assembly varies from one antenna to another, but practically all follow the same pattern. An instruction sheet, outlining a specific method of assembly and adjustment, is generally packed with the antenna. These instructions should be carried out to the letter. It is much easier to spend a little extra time while the antenna is on the ground than to have to take it down later because of a loose element, for example.

Fig. 5-4 is an example of the assembly instructions included with an antenna. This particular illustration is one part of the assembly procedure for a cardioid-pattern

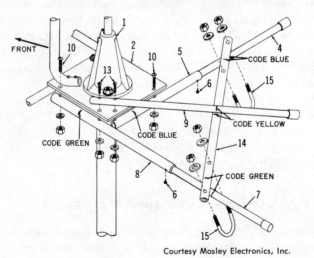

Courtesy Mosley Electronics, Inc.

Fig. 5-4. Assembly details for a typical ground-plane antenna.

ground-plane antenna. Notice that each component is numbered to simplify assembly and prevent errors. The assembly data for a coaxial antenna are shown in Fig. 5-5. The complete assembly is quite simple and involves only what is shown here; the shaded parts have been preassembled at the factory. Most CB antennas can be assembled with little or no trouble, provided the instructions are closely followed.

Antenna Tuning

When an antenna is in tune, its element(s) are of the correct length for optimum response at a given frequency. In this ideal condition, known as *resonance,* the inductive re-

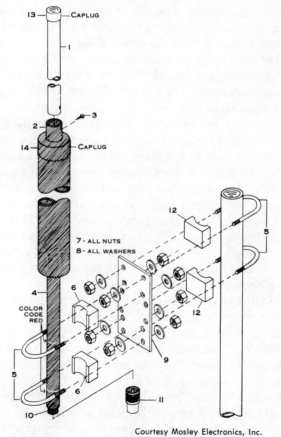

Courtesy Mosley Electronics, Inc.

Fig. 5-5. Assembly and mounting details for a typical coaxial antenna.

actance of the antenna system is equal in value and opposite in polarity to the capacitive reactance. As a result, the two cancel and the antenna appears to the transmitter as a pure resistance—hence, a perfect load. Only under this condition can optimum efficiency be realized. Any antenna is resonant at *some* frequency determined by the length of its elements. The problem is to achieve the ideal (resonant) condition at the desired frequency, and this means that the length of these elements must be correct and any matching devices must be properly adjusted. Only when the antenna is tuned to resonance at the correct frequency will the transmitter deliver maximum power to it.

When adjustments are required, they should be made carefully; otherwise, the antenna may be detuned even further. This, of course, can reduce its efficiency considerably, and if the antenna is detuned too greatly, the final rf amplifier may even be damaged. An improperly tuned antenna presents a reduced load resistance to the final rf stage; consequently, the current in the final rf stage may become excessive.

An antenna is tuned by adjusting or cutting its element(s) to the correct length for a specific frequency. When a matching device is employed, it too must be adjusted. The term "adjustment" is used loosely here; not all antennas are adjustable. A mobile whip, for example, has a fixed length, and if tuning is required, the element must be cut to the correct length. Be careful here—if too much is cut off, you may have to use a coupling device or perhaps even replace the entire rod. Most base-station antennas (ground-plane, coaxial, and beam) have provision for adjusting the lengths of their elements.

Some CB antennas are factory tuned and may be installed without further adjustments. Those that require tuning before being erected are usually accompanied by an instruction sheet outlining the proper procedure. An antenna will operate most efficiently at the resonant frequency. At either side of this frequency, however, its performance will drop off. This can be seen from the swr curves shown previously. The point of lowest swr is the point at or nearest resonance. An antenna to be used for one frequency only should be tuned to resonance at that frequency; however, for operation over a number of frequencies (such as class-D operation), the antenna is generally tuned for resonance at the center of the band and compromise operation is then achieved over the entire band. Another method, although not too practical, involves retuning the antenna for each new operating frequency.

The antenna bandwidth characteristics also affect its performance considerably. The efficiency of some antennas drops off much more rapidly at frequencies on either side of resonance. The swr curve of an antenna gives a good indication of its bandwidth response.

ANTENNA MOUNTING

In class-A operation, the antenna may be mounted on top of a tower or a mast to take full advantage of the height permitted. Tall masts require guying for support, whereas a tower may be self-supporting, guyed, or even both. The installation of towers is a subject within itself and would therefore be impractical to include in a book of this size. Towers are employed mostly with class-A equipment anyway. Occasionally, short towers are used in class-D CB operation. However, since 20 feet is the legal antenna height in this service, masts are usually more practical. This discussion will be concerned primarily with installation of antennas for the class-C and -D services, which are both subject to the 20-foot height limitation.

The way an antenna is mounted depends a great deal on the antenna itself. Some are short enough to be mounted on a mast, which in turn is erected to the proper height and secured. Others are already the legal length and merely require some form of support.

Fig. 5-6 illustrates several methods of mounting a CB antenna. In Fig. 5-6A the antenna is mounted on a mast recessed into the ground and supported by guy wires. The mast in Fig. 5-6B is supported by the side of a building. If the chimney is in good shape, it can serve as a mounting support for the base of the mast, as shown in Fig. 5-6C. The additional support usually required for the upper portion may be provided by guy wires. The CB antenna shown in Fig. 5-6D is self-supporting. Most beam antennas have

(A) Supported by the ground and guy wires.

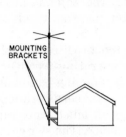

(B) Supported by a building structure.

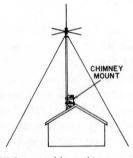

(C) Supported by a chimney and guy wires.

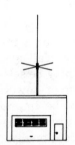

(D) Self-supporting.

Fig. 5-6. Four types of antenna installations.

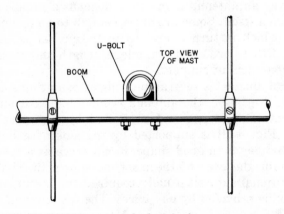

Fig. 5-7. U-bolt arrangement for fastening beam antenna to mast.

U-bolts which pass through the boom (Fig. 5-7) ; the antenna can then be slipped over the top of the mast and secured in place. Some vertical antennas can be secured to the mast as shown in Fig. 5-8.

The type of installation planned will determine the tools and mounting brackets needed. A few of the many brackets available are shown in Fig. 5-9. The one in Fig. 5-9A permits the mast to be fastened to the chimney. The upper portion generally must be supported by guy wires—which are a good idea from the standpoint of safety, even if the chimney does appear sturdy enough. The mounting bracket

Fig. 5-8. One method of securing a vertical antenna to the side of a mast.

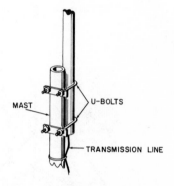

in Fig. 5-9B straddles the cone of the roof. Wood screws are preferred over nails (which pull out more easily) for fastening it down. Waterproof sealing compound should be spread around the screw heads to prevent a leaky roof. A nice feature of this mounting is the swivel socket which permits the mast to be secured while in a horizontal position. Usually the aid of several persons is required. With the antenna and guy wires fastened to the mast and one person steadying each wire, the mast is slipped into the mounting socket and tightened. The mast can now be erected without danger of "loosening it." While one or more helpers (depending on how strong the wind is) steady the raised mast, the slack is taken out of the guy wires and the entire assembly is then secured. A level constructed as shown in Fig. 5-10 eliminates all guesswork in plumbing the mast.

Another type of mounting bracket, shown in Fig. 5-9C, permits the antenna to be mounted on the gable of a build-

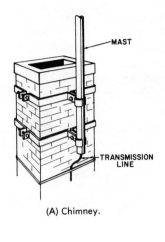

MAST

TRANSMISSION
LINE

(A) Chimney.

(B) Roof cone.

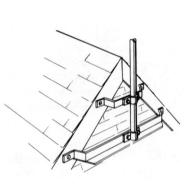

(C) Gable.

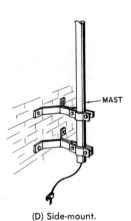

MAST

(D) Side-mount.

Fig. 5-9. Four types of mounting brackets.

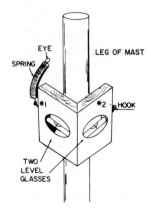

EYE   LEG OF MAST

SPRING

#1   #2   HOOK

TWO
LEVEL
GLASSES

**Fig. 5-10. An easily constructed level
for plumbing masts.**

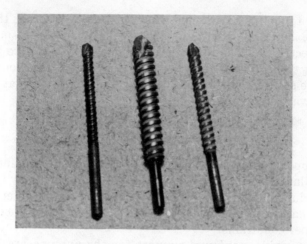

Fig. 5-11. Typical masonry drill bits.

ing. Here again, guy wires may or may not be needed, depending on the condition of the building itself, the type of antenna and its weight, the length of the mast, etc.

In each of the above examples, certain problems may arise that call for a somewhat different approach to the installation methods. The instruction sheet provided with the mounting bracket in Fig. 5-9D, for example, might say to secure this mount to the side of a building with 2-inch wood screws. Yet, many buildings are made of brick and stone. In

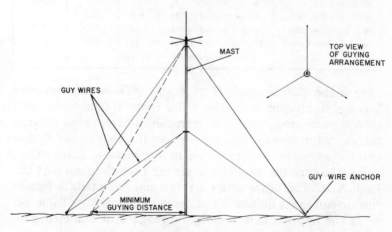

Fig. 5-12. The proper method of guying a mast.

order to mount an antenna mast on the side of other than a frame building, additional tools will be required such as an electric drill and also hard-tip masonry bits like those in Fig. 5-11. After the holes have been drilled, the mounting can be secured with special masonry bolts, available at most hardware stores.

GUYING

Proper guying is also an important part of the antenna installation. If it is done improperly, the antenna may topple over during high winds. Even a properly guyed antenna is not guaranteed to withstand a hurricane or other abnormal weather conditions.

To provide maximum support for the antenna assembly, the guy wires must be properly positioned and secured. A common mistake is anchoring them too close to the base of the mast. Fig. 5-12 shows the proper guying arrangement. The *minimum* safe distance of each guy from the

Table 5-2. Guy-Wire Characteristics

Type	Construction	Strand Diameter (inch)	Approx Breaking Strength (pounds)
Six Strand No. 18	6—Galvanized wires, twisted	.131	640
Six Strand No. 20	6—Galvanized wires, twisted	.095	350
Four Strand No. 20	4—Galvanized wires, twisted	.080	240
No. 3 Cushion Center	11—Galvanized wires, No. 22 around a cushion center	.135	500

base of the mast is one-half its height. (These requirements become significant with the taller-than-20-foot masts and towers permitted in class-A operation.) Also, avoid kinking the guy wire, because this can weaken it considerably at that point. Never use rusty guy wire in a permanent installation, and make sure the guy wire is strong enough to support the load. A stacked beam using a rotor and mounted 20 feet in the air can exert quite a bit of force on guy wires. Table 5-2 gives the approximate breaking strengths of assorted guy-wire sizes.

Just as important as the strength of the guy wire is the anchor and the manner in which the wire is fastened to it. A screw eye is generally used for anchoring to a wooden structure. Make sure the screw is fastened to something solid, such as the framework, and not to weatherboarding or siding, for example, because the guy wire will be only as

Fig. 5-13. The correct method of anchoring and securing guy wires.

strong as the nails holding the material. The anchor should be screwed down as far as possible, and the guy wire should be fastened to it as shown in Fig. 5-13. A grooved metal thimble is slipped through the eye, and the guy wire is then brought around through it and secured. This is done to prevent any sharp bends, which could weaken the guy wire. Fig. 5-14 shows two examples of poor anchoring. Surpris-

Fig. 5-14. Examples of poor guy-wire anchoring.

Fig. 5-15. Typical coaxial cable connectors.

ingly enough, such anchors are quite common—but then, so are battered antennas!

ROUTING AND TERMINATING
THE TRANSMISSION LINE

Before the antenna is erected, the transmission line is connected to the driven element and then neatly dressed

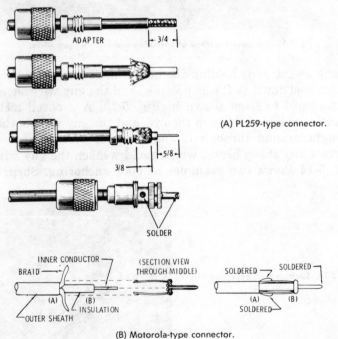

(A) PL259-type connector.

(B) Motorola-type connector.

Courtesy Amphenol Corp.

Fig. 5-16. Two methods of fastening the antenna connector to the coaxial cable.

along the side or through the center of the mast. If the line is run along the outside, it should be either fed through standoff insulators or fastened to the mast with electrical tape, or in some other way prevented from moving around. After the antenna is in place, the transmission line should be carefully routed (by the shortest path) to the radio equipment and secured as required along this path. Any leftover cable at the transceiver should be coiled up rather than cut off. This could preclude splicing later, should the antenna have to be relocated farther away.

The end of the transmission line must now be fitted with the proper coaxial connector so that it can be attached to the radio equipment. Generally this connector will be similar to one of those in Figs. 5-15 or 5-16. Be extremely careful when fastening the connector to the coaxial cable—it is

(A) Light-duty.

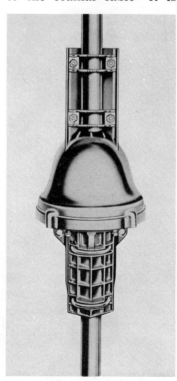

(B) Heavy-duty.

Courtesy Cornell-Dubilier Electronics

Fig. 5-17. Two types of antenna rotors.

easy to short some of the strands of the braid to the center conductor. The correct method of preparing and fastening the cable to the connector is shown in Fig. 5-16. The adapter shown here is used with smaller-diameter coaxial cable (such as RG-58/U) ; it is not needed with RG8/U.

ROTARY-ANTENNA INSTALLATIONS

Since many antennas are directional, some means of turning them must be provided. This can be done either manually or electrically. More than one antenna has been rotated with a pipe wrench or other equally crude make-shift device, but none is as practical as an electric antenna rotor like those in Fig. 5-17. The one in Fig. 5-17A will handle lightweight CB antennas; the heavy-duty rotor in Fig. 5-17B will handle loads up to 150 pounds with no trouble at all.

Stacked arrays, however, are often quite heavy and could place a considerable strain on a light-duty rotor. To take the weight of the antenna assembly off the rotor, a thrust bearing (Fig. 5-18) may be used. As you can see in Fig. 5-19, the mast and thrust bearing now support the weight of the antenna; the only strain on the rotor itself is that required to rotate the antenna assembly.

Rotary-antenna systems are installed much like the simpler systems. There are, however, some additional con-

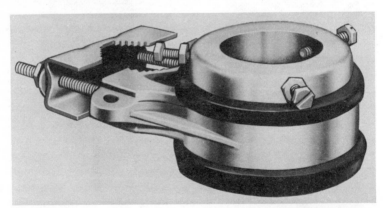

Courtesy Cornell-Dubilier Electronics

Fig. 5-18. A thrust bearing used to take the weight of the antenna off the rotor.

siderations. For one thing, the weight of a rotor often makes guying necessary where it would not be otherwise. Also, there is a rotor control cable which will have to be routed to the rotor control unit.

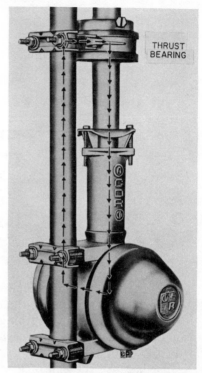

Fig. 5-19. The thrust bearing installed.

Courtesy Cornell-Dubilier Electronics

Allow enough slack in the transmission line, between the points where it is connected to the antenna and secured to the mast, so that the antenna is free to rotate a full 360° without damaging the line, antenna, or rotor. Also, be extremely careful in dressing the rotor cable down the mast and into the control point. A typical rotor installation is shown in Fig. 5-20.

LIGHTNING PROTECTION

No discussion of antenna installation is complete without some mention of lightning protection. Lightning follows the

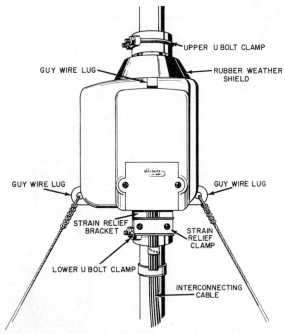

Fig. 5-20. A typical antenna rotor installation.

easiest path to ground, which is often through an antenna. The purpose of grounding the antenna mast is to provide a direct path where the lightning can safely go. Otherwise, it could be hazardous to the operator and may damage or even ruin the radio equipment.

Just because the end of a mast is resting on the ground does not mean that it is properly grounded. A good earth

Fig. 5-21. An in-line lightning-protection device to safeguard both operator and equipment.

Courtesy Hy-Gain Electronics Corp.

ground is required for lightning protection. An outside water faucet makes an ideal ground. If one is not available, a 6-to-9-foot copper ground rod will suffice. The antenna mast should be connected to the ground rod with copper wire of at least 4 to 6 gauge and fastened securely at both ends. Even with the mast properly grounded, there is still the driven element to consider. Many of the newer antenna designs have taken care of this by providing a dc path to the mast for lightning protection. This same path, however, is essentially nonexistent for rf. Obviously, even this method is ineffective unless the mast is properly grounded. Further protection can be provided by inserting the device in Fig. 5-21 in series with the standard PL-259/SO-239 antenna cable connectors. When installed and provided with the proper ground connection, this device protects both operator and radio equipment from lightning and static electrical charges entering on the antenna transmission line.

CHAPTER 6

Mobile Antenna
Installation

Proper selection and installation is as important for a mobile antenna as it is for a base-station antenna. Antenna efficiency is generally of prime importance; however, physical appearance may also have some bearing on the selection of a mobile antenna and the type of installation.

ANTENNA PLACEMENT

Another important consideration is the mounting location of the antenna on the vehicle. The quarter-wave whip uses the metal surface of the vehicle as the ground plane, but different locations present different ground-plane effects, which in turn produce variations in the antenna radiation patterns. Fig. 6-1 shows several mounting locations for mobile antennas. The three in Fig. 6-1 are the most common. Using them as examples, let us see what effect the mounting location has on the radiation pattern. Figs. 6-2, 6-3, and 6-4 show the radiation pattern obtained when a quarter-wave mobile whip is mounted on the roof, rear deck, and bumper, respectively. From these patterns it becomes obvious that the most desirable place to mount the whip is on the roof of the vehicle. However, this is not always possible. For example, a full-length (108-inch) whip would be

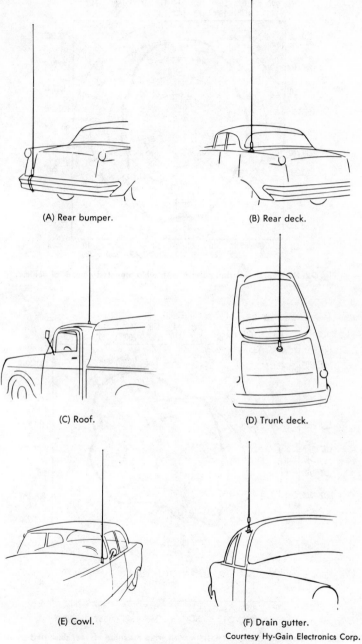

(A) Rear bumper.

(B) Rear deck.

(C) Roof.

(D) Trunk deck.

(E) Cowl.

(F) Drain gutter.

Courtesy Hy-Gain Electronics Corp.

Fig. 6-1. Typical mobile-antenna mounting locations.

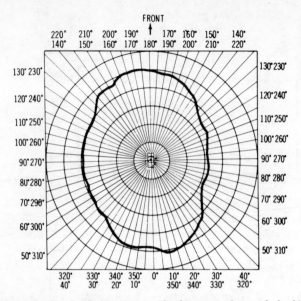

Fig. 6-2. Horizontal radiation pattern with whip mounted on roof of vehicle.

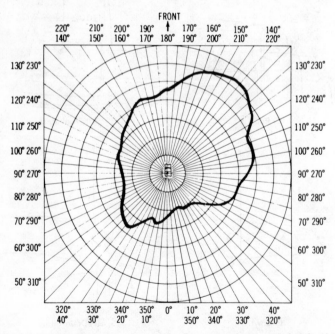

Fig. 6-3. Horizontal radiation pattern with whip mounted on rear deck (left side) of vehicle.

impractical mounted on top of a car or truck. One of the shorter, loaded whips would be more suitable here, but such a change might mean some sacrifice in efficiency. Therefore, the mounting location of the mobile antenna is often a compromise.

MOBILE ANTENNA TYPES

Practically all mobile antennas are of the whip design. Some are a full quarter-wavelength long, while others employ loading coils to permit a reduction in their physical size. Most are constructed of either stainless steel or fiber glass and are available with mountings for permanent or temporary installations.

Problems often associated with mobile antenna installations are minimized by the wide variety of antenna sizes and mounting configurations available. For example, there are mobile antennas equipped with suction-cup magnetic mountings which can be attached to any smooth surface on the vehicle and clamp-on mountings which fasten to the

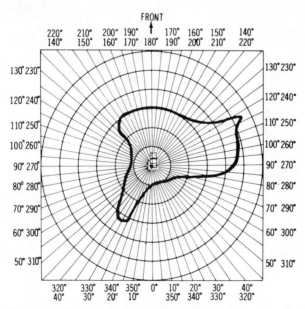

Fig. 6-4. Horizontal radiation pattern with whip mounted on rear bumper (left side) of vehicle.

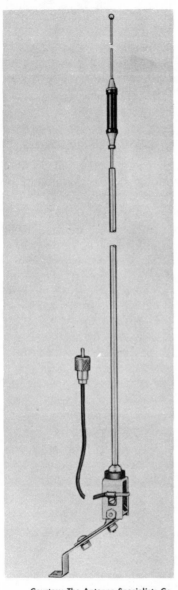

Fig. 6-5. Typical center-loaded mobile whip equipped with trunk-groove mounting.

Fig. 6-6. A center-loaded quarter-wave whip constructed of fiber glass and mounted on the bumper.

rain gutter of the car or truck, or to a luggage rack, the bumper, or the trunk-lid groove. None of these mountings requires drilling holes in the vehicle. There are also special antenna mountings for campers, boats, and other vehicles.

Fig. 6-5 shows a typical center-loaded mobile whip with a mounting that permits the unit to be secured to the trunk-lid groove of a car. In contrast, Fig. 6-6 shows a center-loaded, full quarter-wavelength whip attached to a bumper mount. This whip is constructed of fiber glass (with an embedded conductor) and can be used as a direct replacement for the 108-inch stainless-steel whip. The full quarter-wavelength steel mobile whips usually require a relatively heavy-duty mounting equipped with a spring to absorb the mechanical shocks encountered during vehicular movement. Figs. 6-7 and 6-8 show two types of spring mountings for this purpose. The mounting in Fig. 6-7 swivels and can therefore be mounted on a flat, angular, or vertical surface while maintaining the whip in an upright position. The chain-type mounting in Fig. 6-8 clamps to the bumper and supports the whip in a vertical position. The latter mounting does not require drilling for installation.

INSTALLATION METHODS

The remainder of this chapter covers three of the most popular mobile antenna installations—bumper, rear deck, and roof. Actual installation methods may vary according to the type and manufacturer of the antenna, the peculiarities of the vehicle, and the tools available. For this reason, the procedures discussed are to be considered typical. Specific step-by-step installation instructions are normally provided with the antenna at the time of purchase.

The Bumper Mount

The bumper-mounted mobile antenna is undoubtedly the easiest to install, which probably accounts for much of its popularity. One of the most popular bumper mountings is the chain type which clamps directly over the edges of the bumper, as shown in Fig. 6-8. No hole drilling is required for this type of mounting, and it can be easily adjusted, with the steel mounting links in the center, to fit practically any

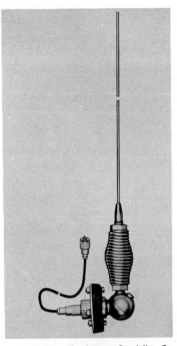 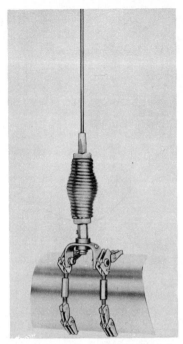

Fig. 6-7. A typical spring-mounted quarter-wave mobile whip.

Fig. 6-8. A popular chain-type bumper mount that can be installed without drilling any holes.

American or foreign car. In Fig. 6-9 you can see how the coaxial transmission line is connected to the mounting. The outer vinyl jacket is removed from that portion of the cable that passes beneath the grounding clip. A suitable terminal lug is soldered to the center conductor, which is then fastened to the base as shown. The whip is then connected to the base, and the cable is routed to the radio equipment. It is better to bring this cable in through the trunk (even though a hole may have to be drilled) and run it beneath the floor-mat to the radio equipment than to route it underneath the car, where it is exposed to any number of hazards.

Rear-Deck and Trunk-Lid Installations

Unless one of the clamp-on types of antenna mounts is used, it will be necessary to drill at least one hole in or near the trunk lid. If a base-loaded antenna with a snap-in mount-

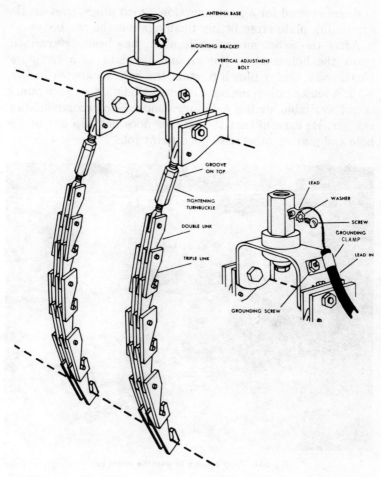

Courtesy The Antenna Specialists Co.

Fig. 6-9. Installation details for the Model M-191 bumper-mount assembly.

ing is to be installed, only one small hole is needed (the exact size will depend on the antenna model), and the installation is relatively simple.

With the spring-mounted swivel base, however, several holes must be drilled in the car body and the installation is somewhat more involved. To install such a mounting on the rear deck of a car, place one hand over the proposed location of the hole. With your other hand, reach inside the trunk (or under the fender with some cars) directly beneath this spot and feel around for any obstruction which might prevent the grounding plate from fitting flush against the car body.

After the exact mounting location has been determined, mark the holes, using the grounding plate as a template. Next, use a center punch to start the hole, as shown in Fig. 6-10. A shear punch makes a clean-cut hole. If a shear punch is not available, drill a hole large enough to accommodate a reamer. Be careful that the reamer does not slip out of the hole and mar an otherwise good paint job.

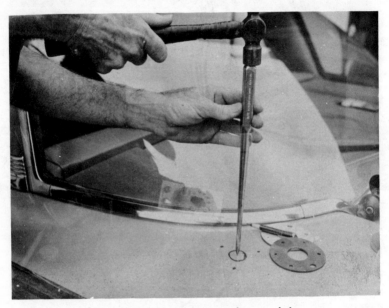

Fig. 6-10. Using a punch to start the center hole.

After all mounting holes have been drilled, assemble the swivel joint and spring mounting, as shown in Fig. 6-11, and mount the assembly in the space provided. Then adjust the swivel joint (Fig. 6-12) until the whip is correctly positioned. Route the transmission line from the transceiver, under the floor mat and rear seat, into the trunk compartment. Then connect the coaxial cable to the base of the whip mounting as shown in Fig. 6-13. Here again, be sure all connections are tight. Notice that a small ferrule is slipped underneath the shielding braid where the cable passes under the grounding clamp. It provides a good contact with the

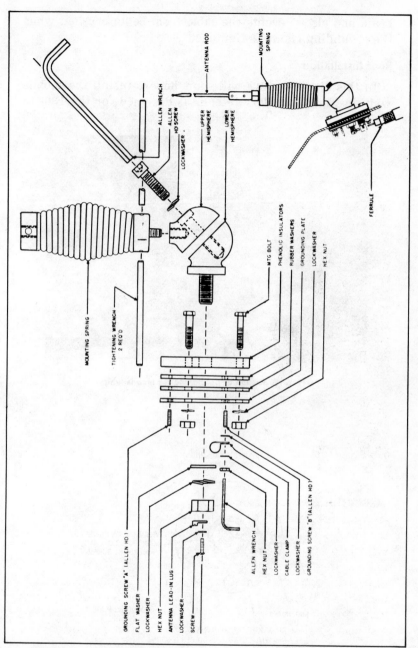

Fig. 6-11. Exploded view of base assembly showing relationship between the
various components.

braid and also prevents the cable from being crushed when the grounding clamp is tightened.

Roof Installation

In this discussion we will show how to install the mobile whip in Fig. 6-14. This antenna is designed for the 130- to

Fig. 6-12. Method of positioning the mobile whip.

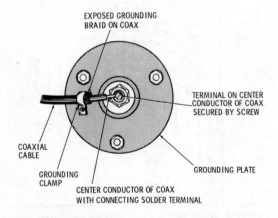

EXPOSED GROUNDING
BRAID ON COAX

TERMINAL ON CENTER
CONDUCTOR OF COAX
SECURED BY SCREW

COAXIAL
CABLE

GROUNDING PLATE

GROUNDING
CLAMP

CENTER CONDUCTOR OF COAX
WITH CONNECTING SOLDER TERMINAL

Fig. 6-13. Method of connecting coaxial cable to swivel base.

470-MHz bands and may therefore be cut for operation at the class-A CB frequencies. It may be mounted on any flat surface, such as the roof, cowl, fender, or rear deck. Here, however, only the roof-top method will be discussed. Mounting the antenna is quite simple. Only a single hole need be drilled, and it is usually unnecessary to remove any upholstery. This same mounting arrangement is used for

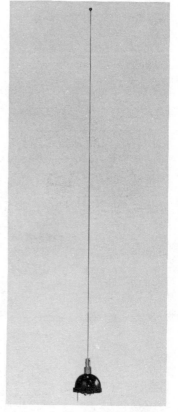

Fig. 6-14. A quarter-wave uhf mobile whip for class-A CB operation.

Courtesy Amphenol Corp.

many of the base-loaded class-D mobile antennas described previously.

The first step is to decide exactly where the antenna should be mounted on the roof. As before, feel around to make sure there are no obstructions, particularly wires leading to the dome light. Now mark, center-punch, and drill

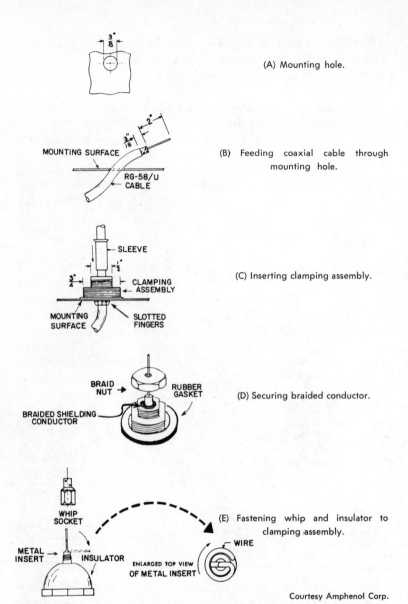

(A) Mounting hole.

(B) Feeding coaxial cable through mounting hole.

(C) Inserting clamping assembly.

(D) Securing braided conductor.

(E) Fastening whip and insulator to clamping assembly.

Courtesy Amphenol Corp.

Fig. 6-15. Method of installing the mobile whip of Fig. 6-14, on the roof of a vehicle.

the proper size hole in the roof (Fig. 6-15A). Feed the coaxial cable through the hole (Fig. 6-15B) and run it between the headliner and the roof to a point where it can be brought out for connection to the radio equipment. This may require the use of a fish tape, or in some instances it may be more practical to drop the headliner. Leave approximately 4 feet of cable exposed and dress the end of it as shown.

Insert the clamping assembly over the exposed end of the cable, as shown in (Fig. 6-15C), and press the slotted fingers (at the base of the clamping assembly) into the hole. Insert the sleeve into the assembly until the flared end is flush with the top. Hold the clamping assembly securely by the ½-inch flats and tighten the threaded nut on the ¾-inch flats until the entire assembly is secured to the mounting surface. Next, fan out the shielding braid, as shown in Fig. 6-15D, and push the cable into the clamping assembly until the braid wires lay flat on top of the assembly.

The inner conductor of the cable should now be extended upward through the braid nut. Then tighten the nut until the braid is securely clamped. Next place the rubber gasket over the clamping assembly and flat against the mounting surface. Extend the inner conductor straight upward through the insulator and metal insert (Fig. 6-15E), and thread the insulator onto the clamping assembly. Pull the wire through the slot of the metal insert and wrap it clockwise around the shoulder of the insert. Any excess wire can now be cut off.

If the whip is not cut to the correct length, it can be removed from its socket by loosening the two small screws. The final step is to thread the whip socket onto the metal insert until the wire is clamped securely against the shoulder.

This completes the roof-antenna installation except for cutting the opposite end of the coaxial cable to the correct length and fastening the proper connector to it. This should be done in the same manner as described previously in Chapter 5.

The exact procedure for mounting any roof antenna may vary not only from one type of antenna to another but also with different manufacturers of the same type of antenna. The procedure described here is intended only as an example of what is normally involved in an installation of this type. In all cases, follow the procedure outlined by the manufacturer for the specific antenna involved.

CHAPTER 7

Improving Existing Antenna Systems

Poor performance in a CB transceiver does not necessarily mean the radio itself is at fault; it could be the antenna system. Granted, a poor antenna will not operate efficiently; but by the same token, even a good antenna will not perform as it should unless the other components in the *system* are doing their job. The antenna "system," you will recall, comprises the antenna, transmission line, and any coupling, loading, or matching devices connected at either end. If even one of the aforementioned components does not perform efficiently, it will adversely affect the whole system. In some instances the inherent characteristics of the radio equipment are such that what might be considered good efficiency can never be achieved. In most installations, however, the efficiency of an existing antenna system can be improved by making certain changes which will be discussed in this chapter. The suggestions offered here will not necessarily apply to every system. For example, if the antenna were already mounted at the maximum height permitted by the FCC, a recommendation to increase antenna height would hardly be appropriate, even though such a change would improve performance.

The most important component in the radiating system is the antenna itself, and therefore it is the most logical place to begin the discussion. Before condemning your present antenna, stop and analyze just what it is designed to do. Perhaps you are expecting too much from this particular antenna or are using the wrong one for the type of communications desired.

RELOCATING THE ANTENNA

The performance of many two-way radio systems can be improved simply by relocating the antenna. This is particularly true for class-A stations. Because of their line-of-sight characteristics, uhf radio waves often provide rather spotty coverage. Where communications are poor between two fixed stations, relocating the antenna 25 to 100 feet from its original location may strengthen the signal substantially. At the class-D frequencies, the change in signal strength will ordinarily be less abrupt since radio signals at these frequencies are reflected much more than they are at uhf.

By making a few comparison measurements with a field-strength meter at the present and proposed antenna locations, you should be able to determine whether or not a move would be advantageous. Since most class-A base stations are relatively permanent installations, these field-strength measurements are generally made preparatory to the initial antenna installation.

The location of an antenna on mobile units also has a considerable effect on performance. Vertical whips are employed in most vehicular installations, and as we saw in the previous chapter, their mounting location has quite an effect on the radiation pattern and hence on antenna performance. The efficiency of the radiating system can be improved in several ways. One method, just mentioned, is to relocate the existing whip at another point where the radiation pattern is improved. Another is to improve the efficiency of the existing antenna or to employ a different one—either replace one of the shorter whips with a full-length quarter-wave job or possibly replace the antenna with one of the new mobile antennas that provide omnidirectional gain. Improving the antenna not only increases the effective radiated power, but also reduces the noise level and minimizes signal fading during reception.

STACKING

Another way of improving the present system is by "stacking" antennas. This involves coupling additional antennas (usually having the same characteristics) to the one being used. One example of a stacked array is shown in Fig. 7-1. As you can see, these antennas may be stacked either horizontally (Fig. 7-1A) or vertically (Fig. 7-1B). Here, an additional three-element beam is mounted beneath or beside the original, and both are connected in parallel to the transmission line.

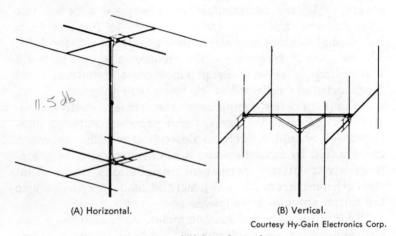

(A) Horizontal.

(B) Vertical.

Courtesy Hy-Gain Electronics Corp.

Fig. 7-1. A stacked three-element beam.

In this example, the single-bay antenna provides a forward gain of 9 dB; but by stacking, an additional 3-dB gain is achieved. This effectively redoubles the power, making a CB transmitter with 5 watts of input equivalent to one having approximately 80 watts.

Most beams are designed so that they can be used either independently or in pairs. So, if stacking is the answer to your problem, a matching antenna, together with an appropriate stacking harness, can usually be obtained from your local electronic parts distributor.

Additions to the existing antenna system need not be in the form of another antenna. They might just as easily be a set of radials added to a standard ground-plane antenna, as shown in Fig. 7-2. In its original state, this antenna provides a low vertical angle of radiation. However, by adding these radials (known as a skirt), the angle of radiation can be lowered still further, thereby making the antenna much more efficient.

INCREASING ANTENNA HEIGHT

One of the best ways of getting more out of an antenna is to increase its height. Unfortunately, the FCC limits CB

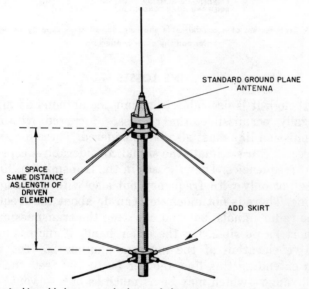

Fig. 7-2. A skirt added to a standard ground-plane antenna to improve its efficiency.

antennas in class-D operation to a height of 20 feet. This is the height of the antenna above its mounting—not the effective elevation above ground. Where the antenna is lower than 20 feet above its mounting, raising it to the maximum limit will usually improve its performance. If it cannot be raised for some reason (e.g., because of overhead obstructions), it is time to consider another mounting location—perhaps a nearby building or other object that will provide more elevation. Remember that the antenna may be any height above ground but only 20 feet above its mounting.

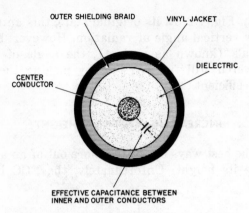

OUTER SHIELDING BRAID   VINYL JACKET

DIELECTRIC

CENTER
CONDUCTOR

EFFECTIVE CAPACITANCE BETWEEN
INNER AND OUTER CONDUCTORS

Fig. 7-3. Cross-section of a coaxial cable showing the effective capacity between the center and the outer conductors.

LINE LOSSES

Although it is desirable to mount the antenna as high as is legally permissible, the rf losses incurred within the transmission line must also be considered. These losses, you will recall, can attenuate the signal considerably—especially at uhf frequencies. The rf losses in the transmission line increase not only with frequency but also with the length of the line. There is not much you can do about the frequency of the radio signals, but you *can* keep the transmission line as short as possible. On the other hand, if increasing the effective elevation of the antenna another 25 feet would mean extending this line another 50 to 100 feet, the additional range—which may be as much as 50%—would overshadow the increased rf loss in the line.

Most rf losses in the coaxial cable are the result of capacitance between the center conductor and the outer shielding braid (Fig. 7-3). One way of minimizing this capacity, and hence the rf losses, is to use a good dielectric material to separate these two conductors. A number of different dielectric materials have been employed, and at a given frequency, some present much less rf loss than others. A reduction in line losses means less signal attenuation and hence better efficiency.

The new RG-8/U or RG-58/U coaxial cable using the foam-type center dielectric is highly recommended because

Table 7-1. Losses Incurred at 27 MHz in Various Antenna Cables

Cable	dB Loss Per 100' @ 27 MHz	% of Power Loss	Wattage at Antenna With 3-Watt Trans. Output		
			100'	200'	300'
RG 58/U	2.6 dB	45%	1.65	.91	.50
RG 8/U	1 dB	20%	2.4	1.92	1.53
½" Foamflex	.44 dB	9%	2.73	2.48	2.26
⅞" Foamflex	.25 dB	5%	2.85	2.71	2.57

of its durability and low loss. Table 7-1 shows how foam dielectric cable compares with regular coaxial cable at a frequency of 27 MHz. A loss of 1 dB in 100 feet of RG-8/U represents a power ratio of 1.25:1, or what amounts to a

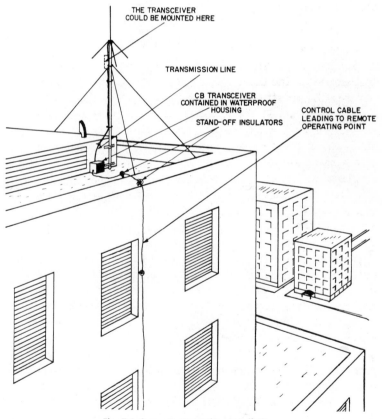

Fig. 7-4. A remote-transceiver installation.

20% power loss due to cable attenuation. If this same cable were 300 feet long, a 3-dB loss would be incurred, which would mean that 50% of the transmitter's output power would never reach the antenna. This being the case, a CB transmitter with 3 watts of output would only deliver 1.5 watts to the antenna, and this is even assuming the antenna is a perfect impedance match. In the ⅞-inch cable using the foam dielectric, the power loss is reduced 15% over the previous type of cable. This would permit the same CB transmitter to deliver 2.5 watts to the antenna. The efficiency of any antenna system, whether for class-A or -D CB operation, can be improved by using a high-quality, low-loss transmission line. Even though it may cost slightly more, chances are you will be more than repaid by the overall improvement in performance.

REMOTE OPERATION

For optimum efficiency, reduce the length of the transmission line as much as possible by mounting the radio equipment at the base or near the top of the antenna tower or mast (Fig. 7-4), and operate it remotely. Sometimes the antenna can be fastened directly to the transceiver, and the transmission line is thereby eliminated altogether! When the radio equipment is to be mounted at the antenna, there is much more freedom in choosing the location that will afford maximum elevation. The reason is that the length of the transmission line is no longer of any great significance. Therefore, in remote installations the efficiency of the antenna system can be improved by (1) eliminating the transmission line and the loss incurred therein, and (2) relocating the antenna at a higher elevation.

CHAPTER **8**

Servicing and Testing Antenna Systems

Troubles in the antenna system can be categorized as physical, electrical, or a combination of both. Some defects occur instantly, while others take awhile to develop. A defect like a broken coax lead is usually readily apparent. On the other hand, one that gradually reduces the efficiency of the antenna system is often hard to recognize and may even go unnoticed for some time. Nevertheless, its degrading effect on the system is still present.

COMMON ANTENNA TROUBLES

The problems encountered in new antenna installations are more likely to be electrical in nature (e.g., improper impedance match), whereas older systems are more prone to exhibit physical ailments. The type of antenna and the service for which it is used (indoor, outdoor, mobile, portable, etc.) also have a bearing on the kinds of troubles most likely to occur. The following discussion is concerned primarily with outdoor antennas; they require the most attention, since they are usually somewhat more complex than the indoor type and are exposed to all the elements of nature.

Guy Wires

Any antenna system is expected to give a certain amount of trouble during its service life, but quite often much of it can be avoided by installing the antenna properly. The strain exerted on an improperly installed guy wire, for example, is often sufficient to break it or pull it loose from its anchor. (This can happen with new as well as old wire.) Or the anchor itself may pull loose from its mounting. Nevertheless, the end result is usually a toppled antenna. When a guy wire—even a new one—is kinked, it is weakened considerably at that point. By examining a broken guy wire closely, you can usually determine why it broke (overloading or kinking), and then you can prevent the same thing from happening to its replacement. Also, any rusted or damaged guy wires should be replaced.

Transmission Lines

Another common source of trouble is the transmission line. Here again, replacement is often necessary because of damages resulting from improper installation. Often an otherwise good cable is damaged because it is not secured or is not routed improperly. An unsecured cable will move about excessively with the wind, and the continuous flexing may break the inner conductor at some point along the line. Sometimes one of the coax leads will break at the point where it is connected to the antenna—a common trouble spot.

There is also the possibility that the transmission line may become worn by rubbing against an object—the edge of a chimney or drain gutter, for example. This may result in a shorted cable. Also, should it be rubbing against a metal object like a gutter, considerable static will probably be noticed during reception. This symptom will be more noticeable on windy days when the line is agitated more violently.

A coaxial cable that is suspected of being shorted or open can be checked out with an ohmmeter by using the methods shown in Fig. 8-1. To check for a short between the center and the outer conductors (Fig. 8-1A), disconnect the cable from the transceiver and place the ohmmeter

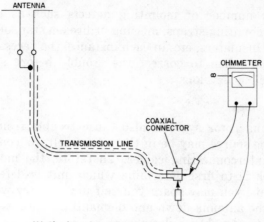

(A) Checking for a short between conductors.

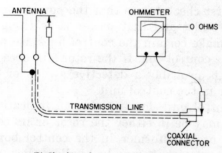

(B) Checking for an open conductor.

Fig. 8-1. Method of checking a coaxial transmission line with an ohmmeter.

probes on the connector as shown. The meter should read infinity. A reading of zero ohms may mean the cable is shorted. This check is not valid, however, for "grounded" antennas which have been designed with a dc path across the line. Here it will be necessary to also disconnect the cable from the antenna before an accurate check can be made. The method, shown in Fig. 8-1B, is used for determining whether or not the line has an open conductor. Broken coax leads at the driven element are sometimes the result of not leaving enough slack in the line when an antenna rotor is employed. The antenna must be able to move freely through 360° of rotation.

Other common physical antenna troubles include loose or corroded antenna connections, missing or damaged elements, and any number of mounting defects such as rusted or broken mounting straps, missing or broken standoff or feed-through insulators, etc. In each instance, the necessary steps should be taken to correct the trouble before something more serious develops.

Rotor Troubles

Antenna rotor systems also cause trouble from time to time. The source may be in the rotor itself, its control unit, or the interconnecting cables. Therefore, the most logical approach is to first determine which unit is defective. By doing this, you may spare yourself unnecessary work (e.g., taking the antenna down and dismantling the rotor, only to find out the trouble is elsewhere).

If the rotor is inoperative and the control unit indicates the same, first check to see that the power cord is plugged into the ac outlet and that power is reaching the control unit. Then make certain the control lines are properly connected to the control box. If the rotor still does not operate, the trouble is probably a defective switch or possibly the transformer in the control unit.

If the control unit does provide an indication, however, the trouble may be in either the rotor or its control. Here, a capacitor or transformer in the control box may be at fault, or the rotor shaft may be binding. Open control leads and open or shorted field coils are also likely possibilities.

An incorrect indication by the control could mean the antenna position does not correspond to the direction indicated, or it could mean a defective indicator, potentiometer, capacitor, or adjustment control. If these check out all right, it is possible that the motor in the rotor is defective or that one of the leads in the control cable is broken.

Sometimes a rotor will rotate in only one direction. Here the trouble may be a control-box switch, or perhaps the rotor has reached its limit of rotation (try reversing).

Another symptom is a rotor that runs slowly or is too weak to turn the antenna. These are common complaints. They can be caused by the motor itself being defective or by a defective capacitor in the control unit. Stripped gears in the rotor are usually indicated when the control unit and motor operate properly, yet the antenna fails to turn.

USING TEST INSTRUMENTS TO CHECK
ANTENNA PERFORMANCE

The quickest and easiest way to check the performance of an antenna system is to use the test instruments designed especially for this purpose.

RF Power Measurements

One of the primary concerns in checking any antenna system is the actual amount of rf power being transferred from the transmitter to the antenna. An rf wattmeter (or powermeter as it is sometimes called) is used to obtain such a measurement.

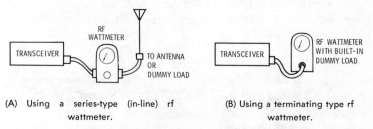

(A) Using a series-type (in-line) rf wattmeter.

(B) Using a terminating type rf wattmeter.

Fig. 8-2. Test setup for measuring transmitter power output.

There are two basic types of rf wattmeters. One is a series, or in-line, type, and the other is a terminating-type meter. The in-line rf wattmeter, as its name implies, is connected in series with the transmission line, between the transmitter output and the antenna (Fig. 8-2A). This type of meter reads the amount of rf power being fed to the antenna. The terminating-type rf wattmeter connects directly to the transmitter output (Fig. 8-2B) and measures the amount of rf power that the transmitter will deliver to a perfectly matched 50-ohm resistive load.

In order to measure output power, it is first necessary to key the transmitter. This should not be done unless the rf energy is fed either to the antenna or to a device known as a dummy load (dummy antenna) which exhibits the proper load impedance. Otherwise, the rf output tube or transistor of the transmitter could be ruined. The dummy load also serves to minimize or eliminate the radiation of radio signals while checking transmitter operation, thereby preventing unnecessary interference to other stations. Fig. 8-3 shows how to construct a simple dummy load using a No. 47 pilot lamp and an antenna connector. Because the lamp glows when rf energy is applied, it can also serve as a visual indication of rf output. A commercial dummy load, like the one in Fig. 8-4, presents a nearly perfect resistive load to the transmitter and at the same time holds down radiation much better.

Whether or not a dummy load is required for power measurements depends on the type of test instruments used. With the in-line type of wattmeter, either a dummy load or the antenna itself must be connected to the output side of the meter (Fig. 8-2A).

The terminating-type wattmeter needs no external device (Fig. 8-2B); it acts as its own load. It can be used for

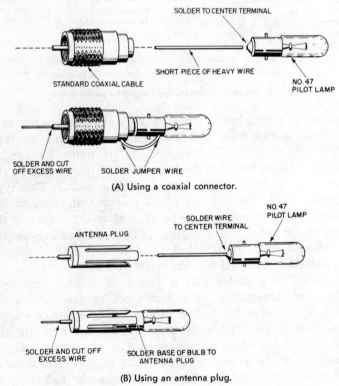

(A) Using a coaxial connector.

(B) Using an antenna plug.

Fig. 8-3. Method of constructing a simple dummy load.

Fig. 8-4. A commercial rf dummy load.

checking transmitter rf output power, but only the in-line type provides an actual measure of the rf power feeding the antenna. Fig. 8-5 shows a typical in-line wattmeter terminated by a dummy load.

A wide variety of rf power indicators is currently available for CB use. Most of these instruments provide "relative rf power output" indications, and some provide other test functions as well.

Fig. 8-5. A typical series-type rf wattmeter terminated with a dummy load.

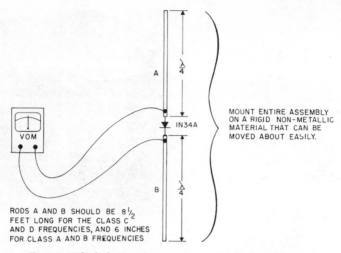

A

$\frac{\lambda}{4}$

1N34A

MOUNT ENTIRE ASSEMBLY
ON A RIGID NON-METALLIC
MATERIAL THAT CAN BE
MOVED ABOUT EASILY.

B

$\frac{\lambda}{4}$

VOM

RODS A AND B SHOULD BE 8$\frac{1}{2}$
FEET LONG FOR THE CLASS C
AND D FREQUENCIES, AND 6 INCHES
FOR CLASS A AND B FREQUENCIES

Fig. 8-6. Method of connecting a vom as an rf field-strength meter.

When an rf meter is not available, a standard vom (volt-ohm-milliammeter) can be used as an rf field-strength meter. This is done by connecting a semiconductor diode and two elements (*A* and *B*), as shown in Fig. 8-6. A relative field-strength indication is obtained by placing the dipole rods a given distance from the transmitting antenna and on the same plane. Naturally, the indication will be useless unless compared with a reference reading that corresponds to the normal field strength at the same distance from the antenna. This method is practical only when the test instrument can be brought close enough to the transmitting

Fig. 8-7. The Seco Model 520A calibrated antenna tester.

Courtesy Seco Electronics, Inc.

antenna to obtain a usable reading (such as with mobile units).

SWR Measurements

Standing-wave ratio measurements are made with an swr bridge. This instrument is a type of in-line rf wattmeter that indicates a power reading in both directions. The face of the meter is calibrated in standing-wave ratios.

The swr bridge is one of the most widely used test instruments for checking the performance of an antenna system. Its operation is based on the fact that a certain amount of the rf energy fed to the antenna is reflected toward the transmitter when the antenna is not properly matched to the transmission line. The antenna acts as a terminating load for this line. If this load is purely resistive and has the same characteristic impedance as the line, the two are said to be matched. Here maximum applied energy is radiated from the antenna; no energy is reflected toward the source, and hence there are no standing waves on the line. Under these ideal conditions, the swr (measure of the mismatch between antenna and line) is 1:1. The swr increases with the amount of mismatch.

The antenna tester shown in Fig. 8-7 reads swr, efficiency, and also relative rf power in watts. An efficient antenna will read within the area labeled GOOD. Directly below the swr scale is another scale that indicates the efficiency of the antenna system. This reading is observed at the same time the swr is checked.

With the transmitter keyed, the relative rf power is determined by turning the FREQUENCY SELECTOR knob (at the right) to indicate the approximate operating frequency and the knob on the left to the FORWARD POWER position. Pressing the POWER X1 button (in the upper right corner) permits the forward power to be read directly from the 10-watt scale. The reflected power is also read from this same scale by turning the selector knob to the setting marked REFLECTED. Suppose the output of a class-D CB transceiver measures 4 watts of forward power. This in itself is good. However, an swr reading of 3:1 would mean the antenna system is only 75% efficient—in other words, ¼ of the forward power is being reflected down the line.

Fig. 8-8. Example of an in-line device designed to reduce or eliminate impedance mismatch.

This represents the loss of 1 watt. The cause of high swr readings (3:1 and above) should be found and corrected if the antenna system is to operate efficiently. High swr can usually be attributed to an antenna or a transmission line with the wrong characteristic impedance; bent, broken, or missing antenna elements; an improperly tuned, constructed, or installed antenna; missing or defective impedance-matching devices (loading coil, etc.); or a damaged transmission line. Even the antenna elevation and surrounding objects affect the swr to some degree. The "CB Matchbox" shown in Fig. 8-8 is an in-line device that can be adjusted to minimize or eliminate most impedance mismatch conditions and thereby improve the swr.

Fig. 8-9 shows an swr bridge using a separate remote indicator. Such an arrangement permits the meter itself to

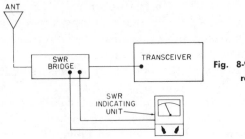

Fig. 8-9. An swr bridge with a remote indicating unit.

be placed at some convenient point in front of the operator while keeping the in-line phasing unit neatly out of sight. This eliminates the need for running the transmission line in front of the radio equipment in order to observe the meter readings. Quite often an swr bridge is left in the line as an rf monitor. It also allows periodic checks of the antenna performance to be made without disturbing other equipment.

CONCLUSION

Much of the trouble encountered in antenna systems can be avoided by careful planning and installation. Particular attention should be given to guying and to the manner in which transmission and control cables are routed and secured along their respective paths. Try to recognize potential troubles and head them off. When trouble does develop, use a logical approach so that you can correct it as quickly and easily as possible without causing further damage. Check the simplest and most likely causes first, and be certain all other avenues have been investigated before finally pulling down the antenna.

One of the best ways to ward off trouble is to check the physical and electrical condition of the antenna system about twice a year—in the spring and again in the fall. A damaged guy wire, for example, found in time, could easily save an antenna. Pay particular attention to the condition of guy wires and their anchors, mounting straps, antenna elements, and connections to the transmission cable. Also check the condition of the transmission line and any rotor control cables, especially at points where they pass through insulators, into buildings, and under and around objects. It is also good preventive maintenance to periodically remove, dismantle, clean and lubricate antenna rotors.

Index